FRENCH SPECIAL FORCES

SPECIAL OPERATIONS COMMAND

Eric Micheletti

HISTOIRE & COLLECTIONS - PARIS

GLOSSARY

ACMS: Civil-Military Affairs

ALAT: French Light Aviation

BACMS: Bureau of Civil-Military Affairs

BAI: Bureau of Influence Operations

BRGE: Electronic Intelligence Brigade

CEMA: Armed Forces Chief of Staff

CEMAT: Army Chief of Staff

CFCA: French Air Force Infantry Command

CIEPCOS: Joint Commission on Special Operations

COFUSCO: Naval Infantry and Special Forces Command

COIA: Joint Operational Center

COMELEF: French Forces Command

COS: Special Operations Command

COSPE: Special Operations Communications

CPA: Air Force Commando Unit

CTM: Maritime Counter Terrorism

CVS: Hover Coupler

DAOS: ALAT Special Operations Detachment

DGA: Directorate General of Acquisitions

DGSE: Directorate of Foreign Operations (French Intelligence Service)

DLDRM: Light Military Intelligence Detachment

DOS: Special Operations Division

DOS-CIET: Special Operations Division-Transport Crew Training Center

DRM: Directorate of Military Intelligence

DSV: HALO

ECT: Command and Communications Section

EHS: Special Forces Helicopter Squadron

ELC: Light Assault Team

EMIA: Joint Staff

EOS : Special Operations Squadron (French Light Aviation helicopters: *ALAT*).

ETRACO: Commando Rapid Transit Asset

FAP: Forward Air Group

FAR: Rapid Reaction Force

FTM: Fardier Helicopter Refueling Vehicle

GCGDA/COS: COS Design Specialist Task Force

GCMA: French-Vietnamese Airborne Commando Group

GCMC: Close Quarters Combat Team

GCP: Airborne Commando Group

GIGN: Special Gendarmerie Assault Unit

GMC: Joint Creativity Group

GSA: Special Autonomous Group

GSIGN: National Gendarmerie Security Group

IHEDN: National Defense Institute

IO: Combat Swimmers (1st RPIMa)

IS: Specialized Training

JVN: Night Vision Goggles

OPSON: Radio Operator

PSY OPS: PSYOPS (Psychological Operations)

RAPAS: Intelligence and Special Operations)

RESCO: CSAR (Combat Search and Rescue)

RTPA: Landing Zone Reconnaissance

SAR: Search and Rescue

SAS: Special Affairs Section

SOTGH: HALO

SPM: Mission Preparation System

SSC: Crisis Resolution Plan

USSOCOM: US Special Forces Command

VRI: Rapid Reconnaissance Vehicle

ISBN: 2 908 182 831

Publisher's number: 2-908182

© Histoire & Collections 1999

5, avenue de la République
F-75541 Paris cedex 11 France
Telephone: 33 1 40 21 18 20
Fax: 33 1 47 00 51 11

This book was designed, assembled and produced by Histoire & Collections on integrated computer systems.
Colour separations: FRT Graphic
Cover design: Patrick Lesieur
Layout: Gil Bourdeaux.
Printed in June 1999
at KSG-Elkar/KSG-Danona, European Union

The United States was the first country to perceive the need to combine Special Forces units under the authority of a permanent joint operational command. Great Britain followed in 1987, and France in 1992.

Other countries already have or will soon follow suit. The logic behind this regrouping of units under a single command is that by doing so countries respond to the need to complement their range of conventional military forces by creating units that are capable of smooth and flexible deployment in times of crisis. It is just this quality of adaptability and rapid reaction that make Special Forces an exceptional tool in the struggle to prevent, limit and control outbreaks of violence in the world. Conflicts today develop without declarations of war and clear demarcation lines between opposing forces are rare; threats come from the most unexpected quarters. Special Forces are the most effective troops for these types of conflicts.

COS units are a part of the best performing military systems in existence today. The Special Forces doctrine is clearly relevant. It is based on two fundamental criteria which only a very few countries have been able to implement: a joint, multi-discipline military structure and extremely high quality unit personnel who have emerged from brutally selective assessment processes and rigorous training programs.

Jacques Saleun
French Air Force General
COS Commander

CONTENTS

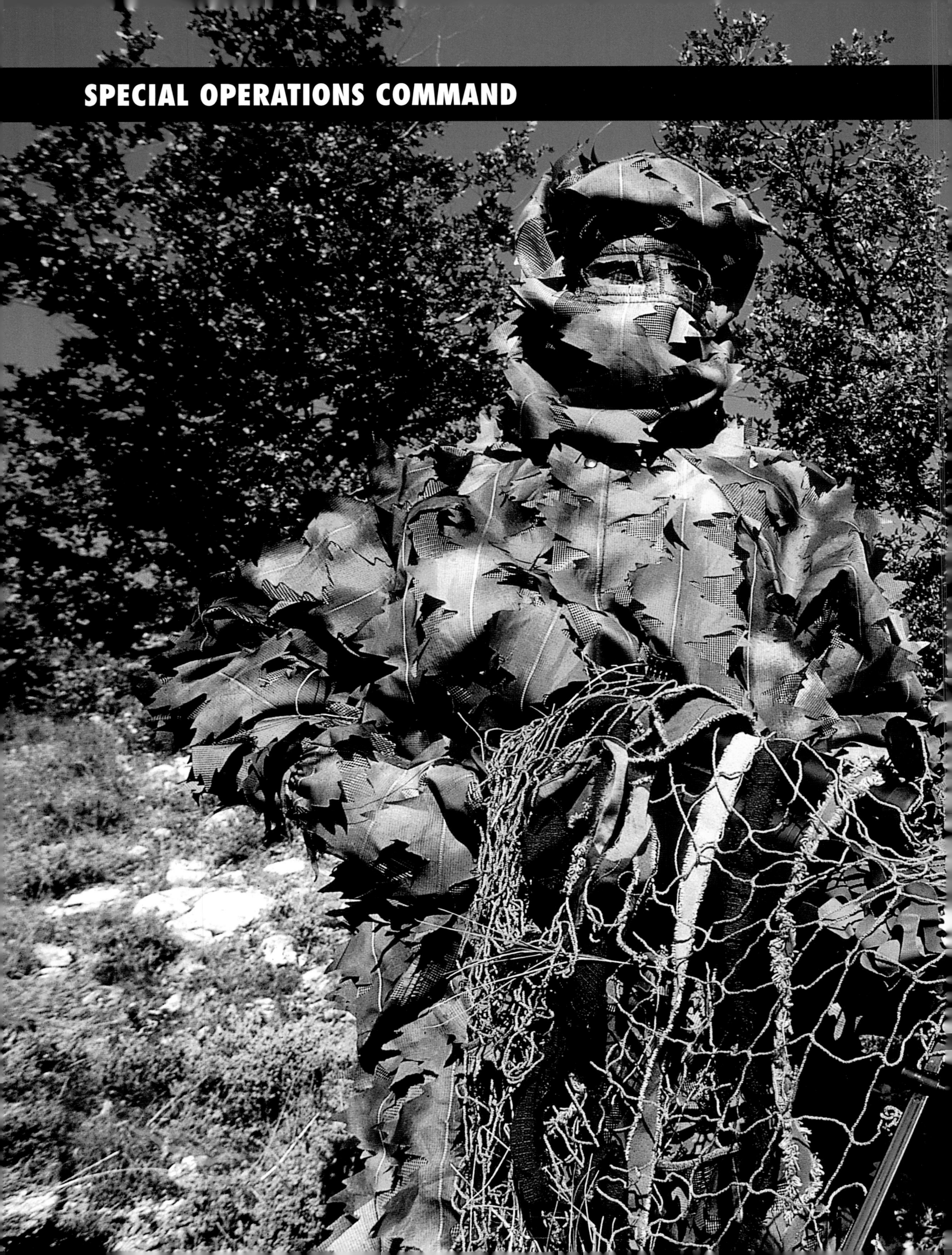

(Commandement des Opérations Spéciales)

The various operations conducted by French units during the Gulf war were replete with learning experiences. One of the most important of these was the need to group Special Forces units of the individual armed services into a joint command structure.

U NLIKE AMERICAN and British Special Forces units that accomplished essential missions in the heart of enemy territory during the war with Iraq, French units were never used to their full extent.

While many assets were available for use at the time—Special Forces units from the three services (French Army, Air Force and Navy), the FAR (Forces d'action rapide—the French version of Rapid Reaction Forces) and the DGSE (*Direction Générale de la Sécurité Extérieure*—French Intelligence Service) units—these did not form a coherent group of military operational assets. It became clear, that for conflicts like the Gulf war, and even smaller operations like those conducted recently in Africa, it was vital to correct certain deficiencies or, in some cases, a complete lack of capability.

The most urgent need was to set up a specific command link between the armed forces chief of staff and Special Forces of the three services, something that previously did not exist. There was also little commonality of equipment and an absence of set operational procedures. Because of these latter problems, commanders were unable to coordinate units on the ground rationally. To this was added a problem of unit organic transportation assets—especially by air—making any in depth penetration impossible. It was also evident that the policy of equipping each service separately resulted in too great a dispersal of finances and made it impossible to acquire special equipment rapidly.

The government and military commanders' acceptance of these problems was the determining factor that led to the creation of a Special Forces command. The new command was placed under the authority of the Armed Forces Chief of Staff and official recognition of Special Forces was secured by a special white paper on defense.

The starting point in the process was in 1992, with the launching of a special commission headed by Major General Le Page. Le Page began by inspecting French Special Forces units to determine their capabilities. Afterwards, he visited the U.S. and Great Britain in order to obtain ideas for designing an appropriate command structure. According to Major General Jacques Saleun, current commander of COS, "The idea was to attain a balance in our Special Forces. They had to be large enough to be taken seriously, capable of joint operations and they needed to have transportation assets of all types at their disposal. We didn't want

Preceding pages: **View of a French Air Force CPA 10 sniper team. COS Special Forces units have four main areas of action: Military assistance, operatioonal support, special direct action (hostage liberation and maritime anti-terrorism) and "influence actions".**

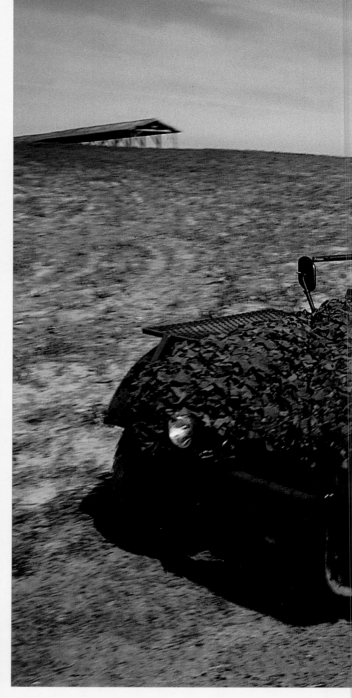

At right:
The rapid reconnaissance vehicle shown here, still in testing, may be chosen for COS units, especcially 1st RPIMa. Special operations missions are approved by the Chief of Staff upon submission by the COS commander.

Below: **COS elements train with a 60mm South African commando mortar in the Jordanian desert. This mortar is the improved, stripped-down model of a similar French version now produced in South Africa. It is very popular because of its simplicity, light weight and speed in setting up.**

Below, at right: **Advance troops of the 1st RPIMa near Gikonoro, Rwanda, during Operation Turquoise in June 1994. COS actions in the country saved many Tsutis from certain death.**

them to be too large; we emphasized quality and training over numbers. After inspecting several special operations commands and groups in the Western hemisphere, we felt that the British concept of about 1,000-man units with joint operations capability was the most appropriate for the French armed forces."

The special operations command that evolved from this effort was given legal status on 24 June 1992. The joint command is under the direct authority of the French Armed Forces Chief of Staff. According to the legislation, its mission is "To plan, coordinate and conduct at the command level all operations carried out by units that are specifically organized, trained and equipped to attain military or paramilitary objectives as defined by the Armed Forces Chief of Staff (CEMA—*Chef d'état-major des armées*)."

To accomplish this mission, the command has at its permanent disposal the so-called elite units. They are available to the command on a per mission basis only, because the units still figure in the organic makeup of their separate services. This elite group, called the premier cercle, is comprised of the foolowing unitsd:

-1st RPIMa (*1er Régiment parachutiste d'infanterie de marine* ; ex-colonial troops with no connection to naval infantry units despite the reference to that branch) and the aviation arm DAOS (*Détachement aérien des opérations spéciales*—Special Operations Aviation Unit), both of whom come under GSA (*Groupement spécial autonome*—Special Independent Command)

-the Naval Commando units Jaubert, de Montfort, de Penfentenyo, Trepel and Hubert with the GCMC (*Groupe de combat en milieu clos*—Close Quarters Combat Team) which is part of COFUSCO (*Commandement des fusiliers-marins commandos*— Naval Infantry and Special Forces Command)

-CPA 10 (*Commando parachutiste de l'air n°10*—Air Force Commando Unit 10)

-EHS : a Special Operations Helicopter Squadron

-DOS : Special Operations Division (Aviation)

In addition, deuxième cercle units apart from this elite group, such as the GIGN (*Groupe d'intervention de la Gendarmerie Nationale*—Special Gendarmerie Assault Unit) of the French Gendarmerie, or other specialized units of the French armed forces may participate in anti-terrorist operations as a part of a COS mission.

The elite Special Forces units are trained to carry out missions in the following four areas: Military assistance, military operations support, counter terrorism and "influence" actions.

Special Forces military assistance missions include training and advising of foreign military personnel abroad, both conventional and special units (i.e., having a specific mission, according to official terminology). This is an especially animated area of activity on the African continent, where many countries have defense treaties with France. The assistance mission also includes participation in humanitarian actions by military personnel. There are

9

THE COS COMMANDER

French Air Force Major General Jacques Saleun was born on 16 March 1945.

In September 1966 he was an air cadet at Salon-de Provence, in the South of France.

Three years later he was assigned to complete course work at the Avord Training Facility.

In March 1970 he was transferred to Airlift Squadron 1/64 in Evreux, moving ahead in his career with a promotion to Aircraft Commander.

After a posting in French Polynesia, he returned to Evreux, successively holding the jobs of Chief of Operations, Deputy Commander and then Commander of the clandestine Air Force Unit GAM 56.

In September 1986 he was admitted to the French War College in Paris, then in 1988 he was promoted to head of Research for the Third Bureau of Air Force Staff.

Following these assignments, he was named Commander of Air Units in Polynesia at Air Base 190 of French South Pacific Forces.

In June 1993, Saleun was designated Chief of Staff at the Special Operations Command at Taverny, after which, following his promotion to Air Force Brigadier General, he took command of COS.

He currently holds the rank of Air Force Lieutenant General.

many examples of French military-led humanitarian operations, ranging from Bosnia to the Congo, all of which had COS detachments as participants.

Another main mission of these units is combat support. Using their specialized skills, Special Forces provide long-range reconnaissance, personal security, night and day air assaults, combat search and rescue missions for downed aviation personnel, unit liaison missions and peacekeeping missions.

The third mission type is specific to commando and counter-terrorist teams and includes resolving hostage incidents, the evacuation of French or allied nationals from foreign territory, long-range operations for field commanders, counter-terrorism oper-

ations at sea (using the Hubert commando unit, GCMC or even the GIGN, with COS approval) and diversion and deception missions.

The last category of COS missions is called actions d'influence or influence operations. This generic term, used to describe what the Americans refer to as "other than war" operations, is gradually gaining acceptance. For COS, the concept means both participation in civil affairs activities and engaging in psychological operations (see the PSYOPS section).

COS detachments are often deployed as deterrence elements on military assistance or support missions. They must be ready on extremely short notice to carry out combat missions like airdrops,

LIEUTENANT GENERAL MAURICE LE PAGE

Lt. General Maurice Le Page was born on 26 January 1939 and was admitted to St. Cyr as a cadet in the school at Coetquidan. He then attended the School of Infantry in Saint-Maixent. In August, 1963 he served as platoon leader in the 6th RPIMa. His next assignment, in 1965, was to the Naval Infantry battalion in Papeete as deputy to the commanding officer, and then later as executive officer to the 3rd Colonial Infantry Regiment in Vannes. In 1970, he started his command time at the rank of Captain with the 3rd RPIMa in Carcassonne. Next he was assigned to the 1st RIAOM (Infantry) in Dakar in 1973, then to a post in the School of Infantry at Montpellier in 1975, after which he enrolled in the War College in 1978. By 1980 he had become deputy commander of the 3rd RPIMa and took part in Operation Barracuda in the Central African Republic as Chief of Staff, and Operation Diodon in Beirut.

In February 1983 he took command of the 8th RPIMa in Castres, the unit that was ordered twice to Chad to take part in Operations Manta and Silure. Two years later he was assigned to the French Army Staff, then in 1988 he became deputy to the commanding general of the military cabinet attached to the Prime Minister. A year later he assumed command of the Airborne Group at Albi.

On 1 December 1991 he was designated representative to CEMA with the job of analyzing the feasibility and the establishment of a Special Forces Command for the three service branches. On 1 August 1992 he was named Commander of Special Operations (COS) and held the job until April 1996. He then became Commander for the Atlantic Defense Military Region, then member of the Army Council until June 1998. On 1 August 1998 General Le Page officially retired from the armed forces.

freeing hostages or long-range operations in enemy territory, then rapidly ressume their primary missions. There are currently several COS detachments deployed outside of France on a permanent basis, carrying out assistance, support and special reconnaissance missions.

The first years following the establishment of COS were spent in overcoming the reluctance of the different services to contribute their Special Forces assets to participate in joint operations, and in achieving the urgent task of lumping together disparate elements into an effective mixture. To accomplish this, a coherent doctrine for the use of Special Forces had to be drawn up, adopting common procedures, writing explicit instruction and training directives and putting into effect an acquisitions policy for units. This had to be accomplished both at the unit level and in concert with the chiefs of staff of the different services. The coordination effort is fundamental because the majority of acquisitions programs are organic to each service.

If the coordination effort was to be effective, the responsibility for its organization would have to fall to the COS office of research and development, charged with heading a joint commission that directs an R & D section and a creativity group for DGA

11

(*Délégation générale d'armement*—Acquisitions Branch)/COS. These two entities will be examined in detail later.

General Saleun, COS commander, states: "The early years were difficult because, through a lack of communication, we were perceived as trouble-makers and competitors. Gradually, through dialogue, presentations and observation of COS units operating in the field, the other chiefs of staff were able to see the need for Special Forces during operations and how each service could use them for specific missions. This way we found our niche. Our experience wasn't unusual: it took nearly ten years for the American command, USSOCOM, to establish itself, and almost fifty years for British Special Forces to attain a foothold in the command staffs of the various services."

General Saluen continues, "My mission is to make available to CEMA what the other units are unable to provide because they are not mission-capable in certain areas. This does not mean that we are in competition with our fellow general staffs—an operation outside of France is a huge joint effort: Special Forces complement conventional forces, they do not compete with them. In civilian evacuations, the presence of Special Forces does not modify missions of other troops in place. Special Forces integrate with units already on the ground, carrying out different types of missions, or missions that require the specialized techniques of these elite units."

COS's big dilemma is urgency. Once a mission is approved, it must be planned very rapidly, in a few hours. COS staff must be capable of drawing up and sending out a draft mission outline, and then determine which units are most appropriate for accomplishing the mission.

The former COS chief of staff outlined the problem: "During the Bangui, Central African Republic operation in 1997, a situation arose requiring immediate reinforcements to in-country units. Several hours after the alert was given the units were already at the air bases where air transport was waiting. But some time constraints, like prepping a Puma for flight operations or flight time to destinations simply cannot be reduced."

Operational methods are firmly established and Special Forces missions are now carried out with very little lag time, except for certain ones requiring full-scale dress rehearsals. During operation Azalée in 1995 command staff wanted units on the ground in four days, but five were taken so that units could practice on a mock up objective.

The COS chief of staff acknowledges, "Whenever CEMA authorizes special reconnaissance for an operation, we don't hesitate. If COS is going to be involved, intelligence is vital. We rely on our own intelligence assets and those of DRM (*Direction de renseignement militaire*— Directorate of Military Intelligence), with whom our units are in direct liaison."

Whenever action appears imminent, a "crisis center" is formed at COS, concentrated entirely on the situation. Should not enough information be available on the problem, DRM or elite unit intelligence assets are called in to help. As the chief of staff asserts, "We consult on everything because we know that good ideas often come from the bottom". During an operation, COS staff becomes a part of the overall system: during operation Pélican in the Congo in 1997, the COS detachment leader, a Lieutenant Colonel, was the deputy of the mission commander, General Landin. In addition, during an operation a COS officer is attached to the COMELEF (*Commandement des éléments français*—French Forces Command): during IFOR missions the French Commanding General was accompanied by a COS Colonel. For these situations a field-grade officer is always present.

Contact is maintained for every operation on a day to day—if not hour to hour—basis, by telecommunications or "live" fax.

All COS units participate in combined exercises, enabling them to operate jointly and to learn certain procedures. There are two every year, one in the spring and one in autumn. The procedures,

At right: A CPA 10 commando during an airport assault exercise. This unit was given the airport assault mission by COS.

At left: A CPA 10 commando with a 60mm mortar in firing position. Although the unit is relatively new and not very large, it rapidly found its place among French Special Forces units through its quality and the missions it accomplished.

Below: COS can use the combat swimmers of the Commando Hubert. It has called on them several times for operations in Africa and Bosnia. Here one of the manned swimmer delivery vehicles about to dive.

12

COS headquarters is comprised of around 50 staff in Taverny (in the Greater Paris region) and in Paris. The HQ has available to it for deployment units with a total of 1,500 elite Special Forces personnel assigned to them. To this may be added the twelve Puma and six Gazelle French Army helicopters with crews belonging to DAOS, amounting to forty or so pilots and specialized mechanics. The French Air Force squadron EHS brings additional Pumas and another 10 or so crew members and thirty more pilots and crew accompany the Air Force C-160 and 130 dedicated transport aircraft available. COS also has high speed rigid and semi-rigid watercraft, swimmer delivery vehicles and a support vessel for divers and their crews.

established over several years, make it possible for Navy elite personnel to train with the EOS (*Escadrille des opérations spéciales*—Special Operations Helicopter Squadron) and the 1st RPIMa to train with the Pyrenees Puma Squadron. The procedures that are practiced are night aircraft landings, landing approaches with night vision goggles, radio transmission procedures, marking drop zones, specialized helicopter-transport and cargo slinging operations, air-dropping equipment and other operations.

The bulk of these exercises take place at night, including 80% of flying missions. Radio transmission and airborne exercises impelled personnel early on to standardize their vocabulary: this is now a matter of routine. Most COS officers will say, "We've come a long way since 1992!"

The best proof of the value of the basic COS concept is that since its creation, no one at the Army Chief of Staff level has found it necessary to modify either the theory governing use of Special Forces or their organization.

France is not currently engaged in any major war effort. Two thirds of all crises that the country responds to involve civil wars elsewhere. These conflicts cannot be settled by either nuclear weapons or the French 1st Army. It was precisely to deal with these new types situations that COS was created. One reserve officer belonging to a civil affairs unit put it this way; "It's a constant struggle involving time and space."

The COS structure is extremely flexible and based on joint services operations. Operating directly under the authority of the Armed Forces Chief of Staff, it allows a very rapid reaction to situations, enabling an equally rapid deployment of all types of Special Forces to anywhere in the world.

COS UNITS IN ACTION

Shortly after the creation of COS (*Commandement des opérations spéciales*—Special Operations Command) in July 1992, the command sent teams to the former Yugoslavia on reconnaissance missions for French forces participating in UN operations.

FROM THAT TIME to the present, COS teams have continued to rotate in and out of the region, especially Bosnia. During the same period, other detachments from the command carried out missions in Africa and in the Comores archipelago of the Indian Ocean. As early as 1992, 1st RPIMa elements not yet under COS authority were part of the 200-man Operation *Osite* sent to these islands following the assassination of the country's President Abdhallah. The objective of this operation was to dislodge mercenaries from an island that they had held for years. This turned out to be a dress rehearsal for an identical operation in October 1995, *Azalée*, with the same opposing force of mercenaries under Bob Denard.

Operation *Oryx* in Somalia took place from December 1992 to February 1993. This military/humanitarian mission was an attempt to restore order and to organize aid distribution to suffering populations. The Taverny staff decided to commit 120 men to the operation, mostly from the 1st RPIMa, but also from the Jaubert and Hubert Naval Commando units and 11th Airborne direct action teams of the 1st and the 35th Airborne Regiments. There were also two 13th Airborne Regiment (Dragons) teams under the command of a French Military Intelligence (DRM) Colonel, who performed long range reconnaissance with COS commandos for other French units.

COS went on alert for this mission on 7 December, and five days later units boarded C-130 Hercules with P-4 reconnaissance vehicles and their gear bound for Djibouti. Upon their arrival on 9 December they found that the Foreign Legion was already in Mogadishu and an analysis of local tendencies had been completed. *Oryx* command staff decided that the best mission for COS units was reconnaissance under the military-humanitarian framework instead of direct action involving taking strategic points in the country. On 14 December, COS elements were integrated into an American column headed for the city of Baido as a part of Operation "Restore Hope". Upon entry into the city, the COS detachment completed numerous reconnaissance missions, one of which

led to the discovery of an ammunitions depot. On 25 December the detachment arrived at Hoddur, where it set up a headquarters. Here the P-4s were exchanged for VLRAs, a vehicle better adapted to desert conditions.

COS elements were transported by Puma to establish surveillance posts on the Ethiopian border to monitor militia infiltration from Ethiopia. In addition to intelligence collection assignments, COS personnel carried out an effective mine and shell disposal mission before returning to France on 12 January 1993.

On 28 January, another COS detachment boarded the military transport *Foudre*, bound for the Adriatic to participate in Operation *Balbuzard*. During 1993, numerous COS teams and detachments were also sent to Africa, successively to Rwanda, Gabon, Congo, Central African Republic, Burundi, the Ivory Coast and Guinea.

Operation *Amaryllis*, conducted in Rwanda from 8-14 April 1994, involved the evacuation of expatriates from Kigali and the surrounding area. France rushed some 500 airborne troops to the Rwanda capital to assist the departure of 1,500 people threatened by the advancing FPR Tutsi forces. The French units involved were 190 members of the 3rd RPIMa stationed in Bangui and a COS detachment made up mostly of 1st RPIMa personnel. They operated with 600 Belgian airborne troops, 90 Italian parachutists of the Folgone Airborne Brigade and 330 U.S Marines held in reserve in Burundi. The operation quickly became difficult, as the local population, made up of a Hutu majority, began killing ethnic Tutsi as FPR elements advanced on the capital. Immediately following the evacuation of last expatriates from Kigali, the operation was shut down and COS commandos were the last to leave.

In fact, COS personnel had already operated in Rwanda. Defense pacts dating from 1975 between France and Rwanda made it natural that the Bayonne-based 1st RPIMa be present in October 1990 for a military assistance mission, dubbed Noroit, involving the DAMI Panda (*Détachement d'assistance militaire et d'instruction*). One aspect of the mission was border surveillance, an exercise named Chimère. In all, 80 troops made up the DAMI, from the 1st RPIMa, the 1st Regiment of Airborne Hussards and the 35th Airborne Artillery Regiment.

In April 1994, two COS operations were conducted in Bosnia. The first consisted in breaking through to the pocket at Goradze encircled by Serb forces and to maintain a 20 kilometer radius neutral zone around Sarajevo and Goradze. The second operation involved an extraction of British SAS from the same pocket by 1st RPIMa troops.

Beginning in early May of the same year,

naval commandos conducted a two week expatriate evacuation operation in Yemen. Then, from 20 June to 9 August, Operation *Turquoise* came up in Rwanda. This operation received intense media coverage at the time, and later in 1998, when a parliamentary inquiry on French policy in Rwanda was initiated. The mission's objective was to create a secure area for refugees and Tutsis that were being threatened with massacre at the hands of the Hutus. COS and Special Gendarmerie units got the word to prepare to ship out to Africa on 17 June. The 150 French troops that ended up going came from the following units:

-1st RPIMa (50 men based in Bayonne)
-*CommandoTrepel* (40 men based in Lorient)
-CPA 10 (based in Nîmes)
-GIGN and EPIGN (based in Satory—Paris region)
-DOS: Transall squadron (1 aircraft, based in Toulouse)
-EHS Pumas (2 rotary aircraft based in Aix-les-Milles)

Preceding pages:
The French government came to terms with mercenaries under Bob Denard in time to avoid hostilities during Operation Azalée in the Comores archipelago from September to October, 1995.

At left, above:
A 1st RPIMa P-4 in Somalia during Operation Oryx mixed into a column of U.S. and COS troops.

At left, above:
COS teams performed numerous in-depth reconnaissance missions as one of the various French detachments participating in Operation Restore Hope in Somalia.

Above:
A COS vehicle, belonging to 1st RPIMa, before a patrol deep into Somalia. Oryx was COS's first opportunity to launch a large-scale deployment of Special Forces units.

Although not officially considered part of the elite units, several long-range commando teams from the 11th Airborne division took part in the operation as well. DRM also sent several 13th RDP Dragon teams to handle reconnaissance and communications missions. They worked in complete harmony with COS personnel.

The commandos loaded their gear into an Airbus, a Hercules, a Transall, an Anotov AN 124 and an Illuyshin IL-76 and headed to Bangui in the Central African Republic (CAR). On 20 June the first Transall loaded with 1st RPIMa troops stationed in Central African Republic (CAR) landed on the tarmac at Goma, Zaire. About 20 troops immediately secured the landing strip in Zaire for the arrival of other Transalls carrying the remainder of the force, a contingent of which would proceed to Bukavu to reconnoiter. The following day the first COS commandos crossed the Zaire-Rwanda border in vehicles and stopped at Cuangugu. That day, Transalls and Hercules began to ferry light P-4 and VLRA reconnaissance vehicles together with support weapons between CAR and Rwanda.

At exactly 1530 hours on 23 June, 46 COS commandos crossed the border in the direction of Cyangugu—officially, this time—inaugurating Operation *Turquoise*. Their mission was to reconnoiter the country, then slow and eventually stop the advancing Tutsis, thereby creating a safe zone for hundreds of thousands of uprooted Rwandans.

COS detachments broke into small teams to engage in reconnaissance sorties of the areas surrounding the cities in southwestern part of the country. The objective of this action was to discover the true intentions of the Tutsi army and to keep the maximum number of Tutsi and Hutu civilians safe from the massacring rampage of extremist Hutus. The five vehicle detachments—four lighter P-4s and a bigger VLRA—fired on FPR Tutsi forces several times during these maneuvers. On other occasions, COS units reconnoitered vast areas going right up to Kigali without encountering a single soldier.

The secure "humanitarian area" eventually established ran from Gikongoro in the South to Kibuye, situated on Lake Kivu. By 28 June all COS vehicles had arrived in Goma, allowing teams to crisscross the region, securing zone *Turquoise* and laying out a "front" interdicting further FPR troop advances. The operation was supported by DOS Transalls and EHS Pumas that kept supplies coming constantly. This was effectively the first Air-Land-Sea mission conducted by COS. After 20 July, four units of COS troops stationed in the region of Gikongoro and supported by 5 Pumas and a C-160 out of Bukavu, were relieved by the RICM (*Régiment d'infanterie et de chars de marine:* Colonial Infantry) and

At least:
COS detachments were deployed for Operation Turquoise in Rwanda, from June to August, 1994, Here a 1st RPIMa commando provides security for the evacuation of Tutsis from the Kibero zone.

At left, below:
COS units got an enthusiastic reception after crossing the Zaire-Rwanda border. The welcome continued as troops progressed West into the country. Here, a member of the Commando Trepel *rides* in a VLRA.

Below:
In the Kibero region a COS detachment, realizing that civilians are in danger in the still active Turquoise zone, attempts to get them out of harm's way.

the last Special Forces troops were pulled out by the beginning of August.

From the summer through October of 1994, infantry and naval detachments were in Haiti to watch over the return of President Aristide and to stand by in case French expatriates needed to be evacuated. Later, from May to July 1995, the *Balbuzard* mission took place in the Adriatic and the former Yugoslavia.

These events were followed by a period of Greenpeace-generated turbulence initiated by the announcement of the resumption of French nuclear testing in the South Pacific. That organization launched an extensive media campaign and sent its ships off to Polynesia. On 8 July, the *Rainbow Warrior* attempted to cross into French territorial waters, prompting the French Navy to launch Operation *Nautile I* to intercept the intruder, an action carried out by the *Commando Hubert*. Although the unit completed its mission, the action spurred indignation worldwide for what was perceived as a very "military" boarding. On 1 September the *Rainbow Warrior* again violated French waters, only to be commandeered by elements of the same unit, this time without masks or weapons. The operation, called *Nautile II*, was reported favorably by the press, with the world observing French military personnel operating in a restrained and courteous fashion.

Operation *Azalée* took place between 30 September and 11 October in the Comoros, with the objective of retaking the island from mercenaries and ex-presidential guard units under Bob Denard.

During the night of 27-28 September, a group of 30 mercenaries disembarked from their boat, the *Vulcain*, and easily took the presidential compound and military camp of Kandani. Their task was facilitated by a receptive local population and by the fact that most of the team already knew the Comoros well from their service in the Presidential Guard. The following day, the French government condemned the invasion and ordered its armed forces onto the island. COS and Indian Ocean units were put on alert immediately. On 1 October, several air transports left loaded with 1st RPIMa troops, the *Commando Jaubert*, *Hubert* combat swimmers, GIGN gendarme special units and EOS Pumas headed for Mayotte. This island served as an operations and launching base for the mission. A part of the assault force boarded two ships, the *Floréal* and the *La Rieuse*, while the other part was to be an air-landed element, using EOS Pumas (one armed with 20 mm cannon).

On 3 October at 2200 hours the *Commando Jaubert* was doing a reconnaissance of the beach directly facing Hahaya airport. At 0200 hours three Pumas brought 1st RPIMa commandos from Mayotte to secure the airstrip

Above:
They managed a coup in the Comoros islands, but now Bob Denard's mercenaries surrender to COS commandos. Several days after the coup, France denounced it and launched Operation Azalée to retake the archipelago.

Center:
1st RPIMa troops and GIGN special police units surround the mercenary camp on 5 October in the pouring rain. The mercenaries surrendered after several hours.

Below:
During the Comoros operation, tense COS commandos shot and killed a Comorian soldier who attempted to pass through a road block. The operation was successful from a military standpoint, but left a disagreeable impression of political duplicity.

At right, above:
1st RPIMa commandos fast-rope from a DAOS Puma following Operation Almadin 2 in Bangui, Central African Republic.

At right, below:
COS commandos escorted expatriates through combat zones during Operation Almadin 2, and later provided security for mutiny leaders during negotiations with the CAR government. Here, mutineers get into a VAB with 1st RPIMa troops.

under blackout, while two other Pumas transporting GIGN personnel took off from the frigate *Floréal* to land at 0230 hours at the island's other airport at Iconi, with the mission of securing the French embassy. Thirty minutes earlier, a team of 13th RDP troops, in company with SPEMER (*Spécialistes mer*— Maritime Operations specialists) teams and 2nd RPIMa commandos landed in Zodiacs on the beaches between Iconi and Mitsoudje. COS personnel advanced rapidly, encountering no resistance. The mercenaries had no cause to fight for, and at any rate, little motivation to combat French troops. By the end of 4 October all objectives had been taken, leaving only the final surrender of Bob Denard and his 33 mercenaries and 400 Camorian soldiers to arrange. This was accomplished by 1458 hours on 5 October.

Operation *Salamandre* was to begin in December 1995. COS anticipated significant activity in the former Yugoslavia in connection with the implementation of the peace plan negotiated for Bosnia. CPA 10 troops stationed in Brindisi, Italy in August of that year were part of a Search and Rescue operation working with American Special Forces during an attempt to recover a Mirage 2000 crew that went down over Serbian-held territory in Bosnia.

May 1996 brought on Operations *Almadin 1* and *2* in the Central African Republic. French troops stationed in the country, including COS forces belonging to the 1st RPIMa, were called in to restore the peace in Bangui after several hundred soldiers and NCOs disgruntled over back pay rebelled and refused to hand over their weapons to the Presidential Guard. France sent in several helicopters to support the troops in Bangui, including a Puma armed with 20 mm cannon. The helicopters were re-supplied by a Transall NG on the ground.

During an ensuing firefight between French forces and the rebels who had occupied the country's communications facility, the Puma fired 20 mm rounds into their midst, inflicting over 100 casualties and effectively ending the rebellion. This stopped short any attempt to overthrow President Patassé, but on 18 June, the mutiny re-ignited and Bangui was soon filled with pillaging rebels. A DGSE (*Direction générale de la sécurité externe*— French intelligence service) aircraft arrived that evening to disembark 11th "*Choc*" personnel who had the job of assisting with the securing of sensitive government documents at the Bangui embassy. They were followed by 1st RPIMa troops who arrived late in the evening. The next day fighting increased between the rebels and the 300-man Presidential Guard force.

By Monday, 20 May there were 1,000 French troops in the Central African Republic, arriving from Chad and Gabon. 1st RPIMa teams worked for two days rounding up 800 French expatriates in Bangui. Meanwhile, pressure was being exerted in Paris for an agreement between the rebels and the Bangui government, and expatriates were evacuated by Special Forces teams aboard Pumas and Transalls. Two days later, the rebel forces tried again to take control of the radio station, but French troops stopped them. COS commandos provided security for rebel leaders during negotiations and finally, on 26 May, agreement was reached and the rebellion ended.

However, by 15 November 1996 a third insurrection had arisen in Bangui. This one was also sparked by troops claiming back pay and fighting broke out again between them and the Presidential Guard. With 2,000 French troops in country—including a COS detachment of 1st RPIMa supported by EOS Pumas—a curfew was imposed on 30 November. This was followed by a month of relative calm, when on 30 December fighting broke out yet again. The assassination of a 6th RPIMa officer and non-commissioned officer by the rebels convinced French commanders of the need for action, and on 4 and 5 January the ordered troops, supported by cannon firing Pumas, into neighborhoods where the rebels had taken refuge. The insurgents were cowed by this action and many of them abandoned the fight. Finally on 24 January, after difficult and protracted negotiations, the third insurrection was officially over. This did not preclude COS troops from remaining in place to assist in the deployment of the pan-African peacekeeping force MINURCA in Bangui.

The months of April to July 1997 were occupied by Operation *Alba* in Albania. On the night of 14 April, the *Commando Jaubert* secured the port of Dürres, in advance of the first VBL armored vehicles belonging to a scouting and reconnaissance platoon that disembarked the following day. Their mission over, the *Commando Jaubert* boarded their ship, *Orage* on 21 April.

Operation *Pélican 1* took place in Brazzaville in 1997. It involved the positioning of French units in the city to oversee the possible evacuation of expatriates living in facing Kinshasa (Zaire). Units involved included the 1st RPIMa and the *Commando de Montfort*. By the end of May the situation appeared to have calmed, and the unit was sent to the gulf of Guinea to participate in Operation *Espadon* in Sierra Leone from 31 May through 4 June, assisting in the evacuation of 737 expatriates by boat from Freetown to Conakry. On 6 June the unit was rushed back to Brazzaville to take part in *Pélican* 2.

Operation *Pélican* 2 began on 16 June 1997 in Brazzaville, Republic of Congo. Clashes

occurred between the Congolese forces and the «Ninja» and «Cobra» militias, resulting in several hundred deaths. The French government launched an operation, dubbing it *Pélican* 2, to evacuate 5,666 expatriates, of which 1,523 were French citizens. Two companies of 2nd REP paratroopers were present to assist in a possible evacuation of expatriates from Kinshasa, Zaire, when the situation suddenly worsened at the beginning of June in Brazzaville. Fighting erupted, and by 9 June there were 1,250 French soldiers from the foreign legion (2nd REP, 1st REC, 2nd REI) as well as the 8th RPIMa and the 68th RA (Artillery). To these were added a COS detachment, a parachute group from the 11th Division Parachutiste and about a hundred troops from the 17th RGP (Engineers). The COS detachment contained personnel from the *Commando Hubert* and *de Montfort*, CPA 10 troops, 1st RPIMa soldiers and EOS Pumas. Each unit's role was clearly outlined and COS combined with 2nd REP teams to carry out numerous sorties in combat areas to round up expatriates who were subsequently airlifted by Transall to Libreville, Gabon. On 16 June the French force began to disassemble.

In 1998 COS detachments took part in several operation in Africa that involved the evacuation of expatriates, including Operations *Iroko* en Guinea Bissau in June and *Malachite* in the Congo in August and October. 1st RPIMa personnel remained in the Congo Republic as a part of the MINURCA force in addition to other French forces stationed in several other African countries under defense agreements.

French Special Forces continue to operate in Bosnia. One COS detachment has been ordered to locate and arrest war criminals, following a request by local government authorities. Operations were accordingly carried out in the southeastern multi-national sector of SFOR that encompasses approximately 30% of Bosnian territory. In January 1999, a Serb indicted by the international tribune for war crimes was stopped at a roadblock set up by 1st RPIMa troops and other French soldiers. The man, whose name was Dragan Gagovic, tried to break through the barrier and was shot to death by French soldiers.

COS units were in a state of constant alert in the early months of 1999, in expectation of a rapid deployment by joint Special Forces including French commandos, British SAS and German KSK elite troops to perform extraction missions for NATO personnel in Kosovo.

These types of missions did not occur, and instead, beginning in February several elite units were "reactivated" to prepare for a possible major deployment to Kosovo to carry out missions as part of the cease-fire accords reached in Rambouillet.

Preceding page, above:
COS units operating in Congo-Brazzaville during Pélican in 1997 included elements of the 1st RPIMa, the Commando Hubert and de Montfort and 13th RDP airborne troops under French Military Intelligence command (DRM). They performed numerous sorties in search of expatriates and set up security for French Military Assistance facilities.

Preceding page, center:
De Montfort and Hubert commandos patrol the Congo river in Zodiacs in the vicinity of Brazzaville, opposite Kinshasa. Originally, Pélican 2 was set up to prepare for possible expatriate evacuations from the Zaire capital city which was being threatened by the advance of troops under Laurent-Désiré Kabila.

Preceding page, bottom:
1st RPIMa troops attempt to maintain order and to prevent further looting— of what little remained to pillage—from a VAB borrowed lent by another French unit on a main Brazzaville thoroughfare in 1997.

At left, above:
The tension remained high throughout Pélican 2, as both belligerent camps strongly criticized the French role during the operation. COS missions in Brazzaville were consequently even more difficult than usual.

At left, center:
COS sent 1st RPIMa troops, de Montfort and Hubert naval commandos and CPA 10 commandos to handle the primary Pélican 2 mission of assembling and protecting expatriate populations up until their departure on French aircraft.

At left, bottom:
1st RPIMa commandos prepare to patrol in a local police vehicle in Brazzaville. Their mission: evaluate the situation and mount a show of force to the local combatants.

Preceding pages:
CPA 10 personnel rappelling from an EHS Puma. Special Operations are approved by Armed Forces Chief of Staff, after recommendation by the COS Commander. Upon approval of an operation, COS issues the operation order for a mission and provides a detachment as rapidly as possible. This detachment acts as the command and control cell for units in the field. It is attached to the command staff in the field but remains under the direct operational control of CEMA.

At left:
CPA 10 Air Force commandos carry out an airport assault exercise, similar to ones undertaken over the last two years in Africa. This type of operation is aimed at preparing existing facilities for equipment or troop landings or for expatriate evacuations.

Next page, from top to bottom: A 1st RPIMa sniper team in the Jordanian desert, where wide open spaces are ideal for this type od training:

Next page, above: An FR-F2 sniping team in action. In foreground, the rifle is equipped with a Texas Instruments thermal camera enabling day or night firing and observation, in clear weather or dense fog.

The Special Operations Command (COS) is a joint operational command under the direct authority of CEMA, the French Armed Forces Chief of Staff. As such, it has only minimal organic autonomy.

COS IS MADE UP OF OFFICERS, NCOs and enlisted personnel based at Taverny, just North of Paris and, of course, on the ground during operations where they act as control elements for tactical Special Forces units.

COS is commanded by a general officer and is the official Special Forces command that groups all armed forces branches under one entity. The commanding officer acts as advisor to the Chief of Staff on all Special Forces operations. He has the authority to conduct certain operations with the approval of the Chief of Staff, and he is responsible for planning and executing joint Special Forces exercises. He cooperates with staff of the other armed services in defining Special Forces use doctrine and he draws up training and instruction directives. He also plans and coordinates joint military exercises with the Special Forces of other countries.

The COS commanding general is assisted by his own chief of staff. As second-in-command, this officer coordinates the activities of the staff specialty chiefs.

COS headquarters is an operational joint entity, prepared to mobilize elements for special reconnaissance missions or to direct Special Forces units during operations.

The COS commander has a field-grade officer representing the various services as a sort of satellite to him, tasked with maintaining communication between COS and that officer's parent service, be it Army, Navy or Air Force. There is also a liaison officer from GSIGN, the lead organization of the French Gendarmerie GIGN. The COS commander has a medical officer attached to his staff to provide support during special operations and specific exercises such as HAHO/HALO jumps and operations using closed circuit respirators for combat swimmers from the *Commando Hubert* and from the 1st RPI-Ma.

Each of these officers attached to the COS commander is considered the official interface with his particular service. The Navy interface offi-

At center:
This shooter is in position with an Hecate II 12.7 mm rifle. He is linked up with an observer through a mini radio station that is feeding him the data he needs for long-range sniping.

At bottom:
This sniper is using the HK MSG 90, a favorite of Special Forces units throughout the world. The weapon is semi-automatic, allowing the shooter accurate, rapid-fire operation. It is also effective for providing security to sniper teams. .

THE "ACTIONS D'INFLUENCE" BUREAU

"Other than war". Modern conflicts are characterized by situations that require new types of action and new weapons for operations. The judicious use of these means can be decisive in the struggle to end hostilities and keep the peace.

In the manner of U.S. Special Forces that are very specialized in MOOTW (Military Operations Other Than War), especially CIMIC (Civil/Military Affairs) and PSY-OPS (Psychological Operations), COS was tasked by CEMA with the mission of acquiring similar capabilities for the French Armed Forces in 1993.

What evolved was the BACMS (*Bureau des affaires civilo-militaires spéciales*— Bureau of Special Civil/Military Affairs). This office has the responsibility of formulating, creating and putting into action

with a view to continuing this tradition of the French soldier as "builder", in every sense of the word. The return in force of civil affairs and psychological operations through COS actions represents an authentic renewal of the will to mold the military machine into a tool that can adapt to modern conflicts, with peace and economic development as a backdrop. Additionally, using "Other than war" assets is an efficient and less costly way out of a crisis, both in terms of personnel and funds.

Reservists were authorized to participate in foreign theaters of operations as of 4 June 1993, when a law was enacted to that effect. Within a year, COS conducted its first foray into Bosnia. Both active duty and reserve officers were deployed to the area. They were special-

reconstruction needs and the establishment of a structure to be called the ICBO, the International Coordination Board.

Less than three days after the inception of the Board, US and British experts landed in the Bosnian capital and began to carry out their own appraisal of the situation. For once, France acted quickly, and by mid-March some 15 reservists who in civilian life are architects, utilities experts, construction managers and industrialists were in country. The French technicians, in coordination with local entrepreneurs, rapidly completed an extremely well researched analysis of local conditions. This was the opportunity to mix in with the local environment and get to know the players.

The analysis was written in Serbo-Croa-

all Other than War activities with a view to facilitating the return of a nation recently engaged in war to peace.

The obvious testing grounds for this type of action were the former Yugoslavia, and to a lesser degree, Rwanda. It became quickly evident that these situations required a major capability on the strategic level that would be an extension of and a complement to armed intervention, involving working on the human, economic, cultural and psychological levels.

The key words for this new type of action were to be "persuasion", "reconciliation" and "reconstruction".

The French Army has no lack experience in this field. An indisputable tradition has been carried over from the country's 19th century colonial experience when the Arab Bureau, the Indigenous Affairs and Special Administrative Sections offices labored to achieve the conquest of hearts and minds so necessary for consummating the final "victory".

French military actions since the end of the war in Algeria have been carried out

ized in economic, cultural and institutional sectors, and thus were soon involved in projects related to security, justice and legislation. Their mission was to evaluate the country's needs and to formulate a plan to meet them, all the while seeking to create openings for French companies during reconsrtuction activities in the local economy. Reserve officers were chosen carefully for this mission, most having close ties with French economic and institutional sectors. The endpoint of this initial action in Bosnia was to provide French plumbing equipment for water purified by the French utility company Lyonnaise des eaux.

In Sarajevo, COS reservists worked closely with Chamber of Commerce directors to establish a French cultural center, which they baptized the André Malraux Center.

The best opportunity for setting up a French civil/military affairs structure in Bosnia came in March 1994, with the passing of UN Security Council resolution 900, that mandated an analysis of Sarajevo

tian, French and English, while the British and American one was only in English. The two analyses were compared and synthesized, then presented later as a basis for locating the funds necessary for reconstruction.

The Anglo-American team was amazed

Above:
In Sarajevo, in July 1994: One of the «Actions d'influence» Bureau officers with the Vice-President of the French Syndicate of Entrepreneurs; Mr. Coppin; and the President of the French construction company Spie-Batignolles, together with the Reconstruction Program Head for SCS (Special Coordination for Sarajevo).

Above, at left:
In August 1994, discussions were held between French business officials and the Bosnian Deputy Minister tasked with the reconstruction of the Neretva Bridge and the railroad linking Ploce, Mostar and Sarajevo. Note the COS officer in uniform, at center.

Next page:
General Soubirou, commander of the 11th French Airborne Division, decorates soldiers in July 1996, including an active duty and two reserve officers from COS, for their actions in Bosnia.

upon reading the report by the study's quality and by the amount of work accomplished by a small French team in a relatively short time.

One outstanding member of this COS reservist group is Major Xavier Guilhou. In country within 48 hours of the accord, he was a member of the team whose mission it was to prepare a report defining the main thrust of French policy as well as to participate in a diplomatic mission in Sarajevo, Mostar and Ploce.

Major Guilhou, who is both a reserve officer and a former IHEDN student (Institut des Hautes Etudes de la Défense Nationale—National Defense Institute; a degree program), is a very active force in making contact with companies and public utilities through military reserve personnel he knows in local municipalities and governments. The French civil/military effort was initially concentrated in Sarajevo, then spread elsewhere in Bosnia and finally to Croatia and Serbia.

"The first thing COS team did," says Col. Doucet, "was to calm the anxieties of the civilian population. EDF crews moved in, operating under fire, to restore electricity. In the meantime, a reserve officer operating as an advisor to the Bosnian president, began drawing up urbanization plans for the end of hostilities. He also created a basis for cultural cooperation.

The next step, to be implemented later, was to secure the major roads in the country, then traffic flows, then the highways and railroad lines. A team was positioned to this end in Mostar (a Reserve officer who works for the French metals group Péchiney), and in Ploce (a marine facilities engineer Reserve officer). Later, the Croatian and Muslim populations asked for help in drawing up first a constitution, then a legal code; this ended with the establishment of a corporate legal code and laws for society as a whole. Croatia adopted the French Maritime Code."

The nomination of Jean Arthuis, another IHEDN man, as France's Finance Minister added a new dimension to the civil affairs mission by ensuring a better follow-through for operations. Still, with few people in France genuinely backing the program, problems persisted. It was dif-

ficult to convince the Foreign Affairs and Finance Ministries, as well as military commanders to overcome their reservations about actions abroad. In fact, two years of dialogue and explanations were needed to modify perceptions about foreign intervention and the risk factors that hitherto stymied effective cooperation in this area.

Fortunately, then CEMA chief Admiral Lanxade, who was a driving force behind the civil affairs concept, kept up the struggle. His successor, General Douin, was another believer in the usefulness of *actions d'influence*. He promoted COS operational effectiveness by ensuring respect of the chain of command among field commanders in Bosnia leery of "civilians" operating in the area. As a result, relations between the COS Bureau and French Forces Commands were excellent throughout the operation. Another factor contributing to the mission's success was the confidence of several French companies that risked financing and equipment in the country before government financial support for the effort had solidified.

To maintain a steady rotation of around twenty reservists operating in Bosnia on actions of influence, recruiting had to be conducted by ORSEM, France's Reserve Forces organization, the

CIVIL AFFAIRS FROM NAPOLEON TO ALGERIA

While it is true that the Americans have for years used Civil Affairs during operations outside of their borders, this idea does not have its roots in the U.S. military, in spite of what is commonly believed.

In fact, the Civil Affairs concept began with the French.

Napoleon Bonaparte was the first to employ this idea in the 18th century. He brought a large number of civilians and military officers who were specialists in Arab history, topography, cartography and the social sciences with him when he set out on his Egyptian campaign.

As a result of their activities, a sort of national awareness evolved among the Egyptian social elite, and traces of France's political and intellectual character remained in the country through the 1960's, despite its long history of British dominance. Thus, for over a century, the Egyptian elite spoke

French and kept in touch with developments in Paris.

This was not, however, the only impact of civil affairs policy in the French empire. The Indigenous Affairs departments in Africa and the Arab Affairs department led by General Lyautey at the close of the 19th century are two clear examples of this. More recently, in Vietnam the renowned GCMA (*Groupement de commandos mixtes aéroportés*—French-Vietnamese Airborne Commando Group), using motivated troops, managed to establish a combat and administrative network behind the Vietminh Army, forcing it to divert thousands of men to put an end to guerilla activities in the rear, troops that could have otherwise pressured French units.

Several years later, in Algeria, the impressive SAS (*Section des affaires spéciales*—Special Affairs Section) experience illus-

trated how the department could significantly reduce the hold the FLN (*Front de Libération National*—National Liberation Front) held over local civilian populations as a part of an effort to help the people living in that country.

To return to the American experience, the U.S. Army Corps of Engineers took over the job of dealing with local populations during the Second World War, adapting it to implement economic and civil affairs programs. The Marshall Plan exemplified this effort by setting up a technical and economic oriented structure.

However, it was the Gulf war that spurred the establishment of a genuine civil-military structure in the U.S. military. Civil Affairs teams moved into Kuwait to gain control over port and oil facilities, thus dominating post-war reconstruction and directing contracts to American firms.

French Navy, IHEDN and through networking. Once in country, French personnel ran into British components operating through the Overseas Development Administration and SAS, as well as U.S. Army Civil Affairs troops and other government or "para-government" organizations. Although twenty times superior in number to the French contingent, the Americans were nonetheless "taken in" by the Frenchies, who they respected greatly and with whom they maintained excellent relations.

The French teams were also nicely touted by the British press, who highlighted their accomplishments to the detriment of the U.K. SAS corps.

COS currently has the files of 600 reservists specialized in various fields that are available to Special Operations actions of influence bureaus. The ideal profile of these ACMS (*Affaires Civilio-Militaires Spéciales*— Civil-Military Affairs) reservists includes having a managerial position in civilian life, being married, stable and with a technical specialty. In emergencies, these officers can be on the ground and performing their mission within several days.

Civil Affairs personnel have used extremely limited resources and huge amounts of positive thinking to achieve spectacular successes in operations conducted in this field over the last five years.

In December 1994 a new mission was inaugurated. Psychological Operations, as was observed in the Gulf, in Bosnia, in Somalia and in Rwanda, is an essential element for combating disinformation.

The new international situation and the development of information technology have brought PSYOPs to center stage,

making it a primary weapon for use by armed forces in times of crisis.

These types of actions were first known as "Special Operational Communication", then " Psychological Actions". The use by US and NATO forces of the term PSY-OPs, as well as the gradual diminishing of the "Algerian" complex, are at the root of Psychological Operations becoming part of French forces terminology. Military forces needed to develop the capacity of making local populations in Bosnia perceive the presence of French troops as reassuring, justifying the authority of UN troops which would confirm the commitment to carry out UN directives and sub-

Above:
COS was able to set up a radio station in Sarajevo (Azur FM) and Mostar (Radio Accord) as a part of its Psychological Operations effort, shown here in 1996. These assets were put in place to combat disinformation being disseminated by former warring factions and to keep local populations informed about SFOR missions.

sequently, those of NATO. The Bossnian experience also showed that it was necessary to update codes of conduct for French troops and to reinforce them.

Actions of influence bureaus attacked the problem, coming up with several action plans which were tested in the Bosnian

environment. COS set up a digital FM band radio station with a twenty kilometers radius run by a Reserve communications officer. In the beginning the station played strictly music. This was followed by sports news, and then general information of local interest, such as bus schedules and hours of operation for aid stations, as a precursor to directing messages to target audiences.

Other PSYOPs methods under consideration included dropping tracts from aircraft or distributing them from vehicles and the use of ground-based or vehicle mounted speakers. Audiovisual equipment can be deployed during meetings in hot areas for briefings on operations underway or as a means of easing tensions and attempting to calm belligerent factions. Only teams of five or six officers or NCOs are needed for these missions (except tract airdrops) and even with reduced numbers they can cover large territories because of their intimate knowledge of the terrain and its political, religious and military leaders. This helps them in getting the message out; on the rebound, they can analyze the situation and adapt the information if needed.

One concrete example of COS PSYOPs activities occurred in 1998 when a CPA 10 team carried out information dissemination missions in Bosnia in an SFOR environment. The message was drawn up by a joint services and joint forces staff, comprised of French, German, British and American civil affairs personnel. Tactical teams, including the CPA 10 unit, then disseminated the information to selected audiences. French Special Forces operated in the area occupied by the Salamandre division.

"ACTIONS OF INFLUENCE": COS RESERVISTS IN BOSNIA

Since March 1994, the hundred-odd reserve officers that have cycled in and out of Bosnia have carried out several types of missions, in addition to the earlier tasks involving reconstruction activities. Their objective was threefold: establish a geographical data base indicating property ownership in the city of Sarajevo, get the airport back to normal

functioning and have special magistrates perform analyses of Bosnian law. As part of attempts to restore a law-based government, hearings were held and expert opinions drawn up aimed at re-establishing functions vital to peace. As a result of French backed economic missions, a French-Bosnian Chamber of Commerce was inaugurated, banking and financial

relations were solidified and numerous other links were developed between local businesses and French companies leading to the establishment of several French corporations. A certain number of reservists with specific job specialties were deployed both in Sarajevo and Mostar to groups charged with local dissemination missions.

Above:
CPA Air Commandos on board an EHS Puma about to execute a series of rappels and pickups. Their deployment to Italy as part of the RESCO effort for NATO aircraft over Bosnia enabled unit members to acquire solid experience in this area.

At right:
View of an ET 2/61st Franche-Comté Hercules based in Orléans. This COS C-130 unit has one aircraft of this type available for use every day, either for a COS mission or for training.

cer is thus the direct contact between the French Navy Staff and COS. COS considers this officer as sort of the gateway to the *Royale* (the French Navy). He functions in that capacity during open water operations, operations that require French Navy helicopters like the Super-Frelon and operations involving submarine deployments. The same applies to the Air Force and Army liaison officers, with regard to the use of their services' troops and equipment.

COS is made up of six Bureaus: Operations, Specialized Training, Research & Development, Telecom-

munications and Information Systems, Civil/Military Action and General Services.

The mission of the Operations Bureau is to gather the greatest amount of intelligence available from other military intelligence groups—mostly from DRM and DGSE—and to deploy teams if needed to draw up feasibility studies for missions. This bureau also has the job of setting up a file for missions in concert with the units needed to carry them out, then to obtain necessary authorizations and to handle relations with the other services.

The Specialized Training Bureau has the mission of coordinating joint Special Forces exercises and exchanges of personnel and maneuvers with Special Forces of other countries, like Jordan, Oman, Tunisia and Morocco. It also sets up training, exer-

cises and deployments by air.

Research and Development organizes two main branches of research. The joint commission on development has the responsibility to carry out analyses on special operations-peculiar equipment, techniques and procedures in liaison with the elite units. The commission also provides technical advice to headquarters of the other services on specific new equipment. The design specialist group DGA/COS is tasked with the same mission but on a larger technological and financial scale, bringing COS officers and DGA technicians together to design, test and produce new materiel.

The Bureau of Telecommunications and Information Systems ensures computer and transmission operations, not only for headquarters units but for DLMO (*Détachement léger militaire outre-mer*—Overseas Military Detachment) elements on the ground.

The Civil/Military Actions Bureau operates the so-called Civil Affairs units that coordinate local information and combat disinformation. These units are composed of reservists with expertise in such civilian occupations as public administration, law, economics, public works and utilities, communication and marketing.

The last of the Bureaus, General Services, ensures the proper functioning of COS, handling administration tasks, finance and other miscellaneous services.

THE BUREAU OF RESEARCH AND DEVELOPMENT

The Equipment Innovation and Development Bureau, one of the six resources at COS's disposition, is tasked with designing and developing new types of equipment for use by Special Forces. To carry out its mission of research and analysis, the Bureau directs the activities of two separate units; the COS/DGA working group staffed by COS personnel and DGA engineers, and the Joint Commission on Operative Research. The working group engineers, tests and produces new hi-tech equipment, while the Commission's mission is to analyze equipment, techniques and specific procedures in coordination with the elite units and as a function of their requirements.

intelligence-collecting mini-drone released by a foot soldier, anti-laser protective glasses, stereo vision for pilots, an indestructible observation balloon for detecting movement in an entire city and an intelligent firing scope with a calculator located in the crosshairs that positions the lens according to various data such as wind, distance, target speed, etc. The Commission reviews more than 100 projects annually, ranging from infrared marking systems for of CPA 10 commandos to waterproof emitter-receivers for the *Commando Hubert* or the scale designed for weighing small equipment to be embarked on board DAOS helicopters.

Above:
Currently being tested by COS units, the Pointer mini-drone intelligence collection system was selected for Special Forces by the R & D Bureau of COS. The system was developed by CAC systems. It has a 10 km range and can be launched by hand and recovered by radio control

At left:
Thomson-CSF-produced Sophie night vision goggles, with the second generation thermal image capacity, were tested and are now used by several Special Forces units.

At left, below:
COS and its R & D services are closely following the development of the FELIN project, which they hope will provide French combat troops with sophisticated communications equipment enabling them to analyze a situation and to recognize and locate the threat.

Below:
Among the many projects under consideration by the R & D Bureau is the individual combat weapon. Here tests are being conducted with the French-designed PAPOP infantry weapon system.

The more important projects being worked on by engineers include night operations equipment such as Night Vision Goggles, precision optics and land—sea—air based secure transmission equipment (mounted or freestanding). The group is also following such avant-garde research developments as the PAPOP 5.56mm/35mm (rifle/grenade launcher) individual combat weapon, a helmet-mounted GPS system, an

Three times yearly priorities are redefined and acquisitions—depending on the amounts, naturally—are made.

1st RPIMa

(1er **Régiment Parachutiste d'Infanterie de Marine)**

The 1st RPIMa embodies especially well the SAS spirit, symbolized by the renowned emblem "*Qui ose gagne*". This unit is the direct descendant of French World War II SAS trained paratroops and has kept up not only the esprit de corps of its predecessors, who distinguished themselves during that war and later in Indochina and in Algeria, but also its mission.

THE COMMANDO philosophy based on small, independent teams permeates 1st RPIMa structure. Unit missions include strategic intelligence collection and direct action well behind the enemy's lines, aimed at weakening strategic points and forcing him to divert resources to protect them.

Because the unit is available for the exclusive use of COS, it no longer is under the authority of either the 11th Airborne Division or the French Rapid Reaction Force, FAR *(Force d'Action Rapide)*, and has thus been able to carve itself a niche in the Army structure. It has not only increased expertise in its vocational area but has expanded its field of operations beyond Africa, long the traditional destination of French Naval Infantry, the branch to which the unit is organically and historically linked. Thus, operations in Europe and the Middle East are now a part of the 1st RPIMa's geographical area.

The regiment is part of the Pau-based GSA command (Groupement Spécial Autonome—Special Autonomous Group), that takes its orders from COS. The unit is oriented toward three main mission types: specialized direct action for operations, intelligence collection and military operational assistance to third countries allied to France through defense pacts.

Although 1st RPIMa organization and structures are well-defined, they

37

1st RPIMa HISTORY

One of the first orders given by General Charles de Gaulle in London on 1 August 1940 was for the formation of a French Free Forces airborne unit.

In November of that year, the first elements of a new unit christened *1re Compagnie d'infanterie de l'air* (1st Airborne Infantry Company) reported for airborne training at the Ringway jump school facility. The following year, in the month of March, five French parachutists were dropped in the Morbihan region of France with the mission—code-named Savannah—of attacking a Luftwaffe unit.

After this time, the company was split into two groups. The first was put at the disposition of Special Operations Executive, while the second, under the name of Peloton parachutiste du Levant (Levant Airborne Section), left to take part in operations in the Middle East. This unit was attached to the U.K. Special Air Service Brigade and subsequently was re-named *1re Compagnie de chasseurs parachutistes* (1st Airborne Infantry Company). The origins of the 1st RPIMa motto can be traced to this time: *"Qui ose gagne"* is the translation of the SAS devise.

The "French Squadron" operated extensively in North Africa and the Eastern Mediterranean behind Axis lines, picking up several campaign ribbons along the way, including "Crete 1942", "Libya 1942" and "Southern Tunisia 1942".

Upon their return to Scotland, the two companies were joined to form the *1er battallion d'infanterie de l'air* (1st Airborne Infantry Battalion) whose name was to be changed a few days before D-Day to *2e Régiment de chasseurs parachutistes* (*2e RCP*—2nd Airborne Infantry Regiment).

Airborne troops from this unit landed in Brittany on 6 June with the mission of harassing German troops stationed in the region. The unit drove to the Loire river in armed jeeps, then continued to reap havoc among Germans retreating through the Champagne country. The regiment was sent to the Ardennes in 1944, then to Holland in 1945, before being deactivated at the end of the war.

However, the respite from war was short-lived: the Indochina campaign began gearing up by 1946, and the French government responded by forming up a large contingent of airborne troops. In February 1946, a unit to be called the 1st Airborne Battalion was formed, followed by a second one, ending up in the establishment of a unit of one half brigade strength that was called the *1re demi-brigade de parachutistes SAS*, which continued in the mold of the former *2^e RCP*. Following several changes and

reorganizations a unit called the *1re demi-brigade coloniale de commando parachutistes* came into being, comprising a headquarters, a headquarters battalion and three colonial battalions. After this, the unit's organization remained unchanged through 1954. It took part in all major actions of the war in Indochina, including Hanoi (1946), Nam Dim (1950), Dong Khe (1950), Mao Khe Day (1951), Thai Binh Tule (1952), Na San (1953) and Dien Bien Phu (1954).

The war in Indochina was winding down by

Above:
Shown here a year before the creation of COS, this 1st RPIMa long range reconnaissance team, now called GCPs (Groupement de commandos parachutistes—Airborne Commando Group) had the job of reconnaissance for the Daguet Division during the Gulf war.

Above:
1st RPIMa teams provide security for rebel leaders during negotiations with the Central African Republic government in Bangui, during Operation Almadin 2 in May 1996.

Opposite page:
The first deployment of several COS units was in 1992, for Operation Oryx in Somalia. 1st RPIMa sent the largest detachment.

At right:
1st RPIMa « bérets rouges » oversee the departure of refugees during the 1994 Operation Turquoise in Rwanda. COS units were among the first French troops deployed for this operation, which was later the subject of much discussion.

the end of 1954, while another one was just beginning in Algeria. By 1955, the half-Brigade gave way to the full size colonial Airborne Brigade unit that commanders had envisioned in the beginning. It was made up of a headquarters section, a training section and two Airborne groups (*GAP—Groupement Aéroportée*), and it set up in Bayonne. The Brigade grew rapidly to accommodate the increased demand for troops, and by January 1956, it comprised a headquarters and two regiments, the 2nd and 3rd Colonial Airborne Regiments. It also served as a transit center for troops in training. In 1958, the Brigade was designated Training Brigade for Colonial Airborne Troops, after which it became the Overseas Airborne Brigade and then, in 1960 the Naval Airborne Infantry Brigade. Finally, in October 1962, the unit was dissolved to give way to the 1e Régiment Parachutiste d'Infanterie de marine, or the 1st RPIMa.

Beginning in the seventies, 1st RPIMa troops began taking part in the major deployments that were occurring in Africa. In Zaire, 1st RPIMa elements performed reconnaissance missions on rebels coming from Angola in 1977. Then parts of the unit participated in Operation Verveine in 1979 in the Central African Republic and in Operation Barracuda aimed at ousting the emperor Boukassa in favor of David Dacko. During this operation the 1st RPIMa action group was divided into four detachments, each with the mission of taking official "palace" buildings in the ruler's complex, a mission they performed with exceptional competence.

In August 1983, at the request of Chadian President Habré, 1st RPIMa troops were officially the first French forces—not counting mercenaries and the French intelligence arm "Service action" who directed them in opposing the Libyans —to disembark in the country. The Operation was dubbed Manta, and put in place 4,000 troops to stop the advance of Libyan armed forces. RPIMa troops remained in the country afterward to serve as instructors for the Chadian army. This operation, called Epervier, involved 900 men and continues today.

From May to June 1990, elements of the 1st RPIMa took part in Operation Requin, in Gabon, as part of a 2,000-man French force with the mission of protecting expatriates in the wake of riots in Port-Gentil and Libreville.

A CRAP unit (*Commando de Recherche et d'Action dans la Profondeur*—Long Range Reconnaissance and Direct Action Unit) was formed during the Gulf war, made up of 10 two-man teams and a twenty member headquarters section. The group operated under a 1st RPIMa regimental command and had the mission of reconnoitering for the Daguet division and the U.S. 82nd Airborne division. Two Bayonne men were killed in action at the Al Salman fort during this operation. In May 1991 the regiment got a third palm decoration for the part its members played in the Daguet operation.

A little known mission came up in 1991 in Togo. Christened Verdier, it involved deploying 450 men, including a 1st RPIMa continent, to neighboring Benin following a coup attempt against Togolese transition Premier Kokou Koffigah. At the end of 1991 a mission similar to Requin came up in Zaire involving French and Belgian forces. It lasted through early 1992, with the Bayonne-based unit protecting and evacuating expatriates

following turbulence in Kinshasa. Later in 1992, the unit took part for the first time in a mission administered by the newly created COS in Somalia, called operation Oryx. This operation was carried out as a part of the U.S. Restore Hope operation, with the 1st RPIMa contingent of 120 men led by then Colonel Rozier—since promoted to General and currently GSA commander-deploying in intelligence collection forays. Still in 1992, the unit was part of a 200-man force participating in Operation Osite in the Comores islands following the assassination of President Abdallah. The mission was to remove the mercenaries that had held the island for years.

In accordance with defense pacts signed in 1975 with Rwanda, the 1st RPIMa was sent to that country early on, beginning in October 1990 and carrying through to December 1993, for operation Noroît. This action was aimed at assisting Rwandan armed forces by means of a DAMI (*Détachement d'Assistance Militaire et d'Instruction*—Military Advisory and Instruction Detachment) called Panda. The unit was also called upon to provide surveillance of the borders, a mission formalized under the name of Operation Chimère. There were around 80 troops that made up the DAMI, the 1st RPIMa, the *1er Régiment de hussards parachutistes* and the 35th RAP Airborne Regiment.

In April 1994, the 1st RPIMa returned to Rwanda with other naval troops for Amaryllis with the mission of evacuating expatriates from Kigali and the surrounding area. Then from 22 June to 22 August Operation Turquoise took place, involv-

ing the 1st RPIMa, the Naval Commando units and CPA 10 with EHS providing helicopter support and DOS the airlifting capability.

By October 1995 mercenaries and rebel troops had returned to power in the Comores, sparking Operation Azalée, undertaken to disarm them. This was accomplished in a few days, again using a Special Forces-centered operation run by COS and using 1st RPIMa units, GIGN, the Naval Commandos units, EOS and DOS.

May 1996 saw a deployment by the 1st RPIMa to Bangui, Central African Republic, to carry out Operation Almadin 2, aimed at restoring order to the capital. Evidently the problems were not entirely resolved between the government and rebel troops, for COS commandos returned in December to carry out Operation Almadin 2, Phase II. This operation lasted until January 1997 and involved direct action against the rebels by elements to the 1st RPIMa supported by EOS Pumas.

In March 1997 Operation Pélican 1 started up in Congo-Brazzaville. Units were sent to the country in anticipation of emergency evacuations of Westerners living in Kinshasa in the face of possible violence resulting from forces in rebellion against President Mobutu that were driving toward the capital of Zaire at the time. In reality, the situation exploded in June in Congo-Brazzaville, bringing on Operation Pélican 2. The COS was naturally involved, with 1st RPIMa units sent in to reconnoiter the Congolese capital and to round up Western expatriates that had been isolated in the war torn city.

Above:
A 1st RPIMa paratrooper provides security for other vehicles driven by COS personnel from a crossroads in Brazzaville during Operation Pélican 2. He is armed with an M-16/733, equipped with a 40 mm grenade launcher.

At right, above:
1st RPIMa troops perform motorized reconnaissance in search of expatriates in and around Brazzaville from a VAB and civilian vehicles during Operation Pélican 2 in June 1997.

At right, below:
View of a RAPAS sniper with a 7.62 mm silencer-integrated Ultima Ratio heavy sniper rifle.

may be adapted to fit particular missions. For example, if a Corps commander needs a combination of differently trained personnel--say, a secure facilities acess specialist, a sniper team and combat divers-- for a certain operation, he'll select appropriate personnel from one of the three RAPAS companies of the regiment.

The RAPAS squad (*Recherche et Action Spécialisée*—Intelligence and Special Operations) is the operational nucleus of the regiment. It is made up of 10 commandos, led by an experienced officer or senior NCO and is further broken down into three 2-man teams and a command cell, made up of four persons: a team leader, a radio operator, a communications specialist and a medic. The team leader usually has the "Foreign Area Intelligence" specialty. Since RAPAS teams are completely independent, they are equipped with advanced, long range communications gear and are armed to repel a substantial enemy threat. Each two-man team is specialized in a particular area such as sabotage, accessing secure facilities for military objectives, climbing, fast-roping, wall scal-ing, or shooting.

As explained, RAPAS groups can be assembled from disparate regimental elements to comprise a larger scale, adaptable detachment which is then called an operations and intelligence group, and which can comprise up to 150 or 160 personnel.

The 1st RPIMa has a regimental headquarters, a command and general services company, a training company (considered a RAPAS company as well during peacetime) and three RAPAS companies. These latter three components are considered the lead units of the regiment during specific operations. 1st Company handles non-urban direct action operations, water crossings and water environment missions, as well as personal security for

Above, below and at right:
Views of various phases of preparation for a jump on the tarmac just prior to a training mission. For several years now, GSA has been trying to harmonize operational procedures by using DOS equipment as frequently as possible.

dignitaries. 2nd Company is an urban warfare unit, and also specializes in explosives, sabotage, secure facilities access and sniper operations. 3rd Company provides fire support with heavy mortars, and provides air defense protection. It also specializes in long-range, light vehicle reconnaissance missions after the SAS model.

CCS is comprised of three GCPs (*Groupement de Commandos Parachutiste*—Airborne Commando Group) in addition to the headquarters section and the Regimental Command post. GCP teams not only have the RAPAS qualification, but are also HAHO/HALO qualified and several team members possess the tandem specialty.

There is also an intelligence section that manages information on numerous countries and armies, and is responsible for the regiment's intelligence training. Another cell, called the specialized training cell, works with new equipment and techniques to be adapted to Special Forces missions.

Only active duty soldiers can volunteer to serve in the unit. Officers and cadre come mostly from Naval Infantry units. However, cadre from other branches sometimes turn up in the 1st RPIMa, especially from mountain or cavalry units. These soldiers contribute their share of professional experience, expertise and qualifications. The unit has an exceptionally high cadre to enlisted ratio: more than 30% of its members are officers or NCOs.

The selection process for entry into the 1st RPIMa begins with a review of the applicant's file. General academic performance must be a minimum of 12 out 20, above average by French standards. The review is followed by psychological tests. If the candidate successfully completes these phases he will be admitted, on a space available basis, to a nine month training

THE VLRA (4 x 4) VÉHICULE LÉGER DE RECONNAISSANCE ET D'APPUI— LIGHT RECONNAISSANCE AND SUPPORT VEHICLE

The company ACMAT has developed a heavily armed version of its 4 x 4 VLRA for the 1st RPIMa. The VLRA ALM TPK 4.20 STL.2 has four mounts for .50 caliber, 7.62 AA 52 and 5.56 Minimi machineguns, as well as for 40 mm automatic grenade and smoke grenade launchers. Sand extrication plates, when not in use, can be rotated down to make shelves on the sides of the vehicle, giving additional personal gear carrying capacity during transit.

This vehicle will be used by reconnaissance teams. It will be used as either a troop transport, carrying 12 men with their gear including the driver and the vehicle commander, or as a mule for the rapid reconnaissance vehicles that the unit is expecting to receive shortly.

For now, a tactical setup is being employed using the old P-4s in place of the quick reconnaissance vehicles. Five light vehicles range out front, while two heavier ACMATs trail, handling the transportation of reserve teams and equipment (food, ammunition, personal gear, etc.) The ACMAT has made a good impression on the British SAS, who are replacing their aging Unimogs with ACMAT VLRA ALM TPL 4.20 STL.2s.

The 1st RPIMa also has a number of other light vehicles in its pool besides the new model.

VLRA TECHNICAL DATA

Length: 5.94 m
Width: 2.10 m
Height: 2.26 m (without side rails)
Empty weight: 5 tons
Loaded: 7.5 tons
Gradient capacity: 65% slope
Fording: to 90 cm
Range: 1,600 km (34 hours continuous use)
Motor: Diesel; 6 cylinder Perkins; 138 hp, 2,800 rpm
Gearing: 4 forward, 1 reverse (8 total)
Transfer case: 2WD and 4WD, with differential lock
Suspension (front and rear): coil springs with telescopic shock absorbers
Fuel tank: 180 liters
Water tank (potable): 200 liters
Maximum speed: 100 kpm; 50 kpm with reduction gear

At right:
View of the ACMAT VLRA, recently modified for« heavy » weapons, used by long-range motorized reconnaissance teams. The vehicle boasts four light AA-52 7.62 mm machine guns—three of which have night vision sights—and on 12.7 mm heavy machine gun. This vehicle can transport ten men and all their equipment.

one of the RAPAS companies where he learns the section's specialty. During his second year he may be offered the chance to obtain the next level certificate, CM-1, and eventually, the CT1 RAPAS qualification. This is an in-depth, 15-month advanced commando course offered by the Regiment that helps participants acquire and perfect intelligence gathering skills. This candidate's objective for the course is to obtain the CT1 RAPAS and the CT100 rating. The course also concentrates

THE "RAPIÈRE" COURSE

Everybody, officer and enlisted, is initiated into the unit by taking part in the Rapière (Rapier) course.

The objective of this exercise is to evaluate new members' physical stamina and motivation.

Since all members, regardless of rank, take part in the same tests unit cohesion is strengthened.

One officer says, "This initial training course provides the basis for our intelligence and direct action techniques."

phase. The attrition rate of this course is greater than 50 %. Candidates who successfully complete the course are subsequently offered a three year contract with the unit.

The nine month period is spent at the Regimental training facility. The first two months are dedicated to basic training and airborne qualification. The recruit then pursues the CME—Certificat militaire élémentaire—or, Basic Military Training Certificate, followed by the Level I commando course. After completing these phases, he is assigned an appropriate specialty, usually either RAPAS team member or communications specialist. Only when he completes this training is he assigned to

heavily on Level 2 commando skills. A senior corporal with the CT1 RAPAS rating may next advance to the non-commissioned officer school, which accepts about 30 candidates per year.

The 1st RPIMa has a close historical link with the British Special Air Services. The French unit conducts training with that group in Great Britain and in France, and also holds exercises jointly with the U.S. Army Special Forces.

RAPAS groups regularly conduct jungle training in Guyana out of the COS camp at Régina, and desert exercises in Djibouti. The regiment has recently begun urban combat instruction at a special facility in Pau, which was built nearby the GSA command

Above:
These 1st RPIMa troops from the specialized training company perform a " grappe " exercise from an EOS Puma during a demonstration of the unit's expertise.

facility. Even before becoming one of the elite circle units of the COS, the 1st RPIMa attempted to develop specific capabilities. The regiment worked in a maritime setting, training several groups in nautical operations, and it provided instruction to several groups in counter-terrorism using GIGN methods. The unit also took on the personal security mission, protecting French and foreign general officers on visits to the field. Assignments of this type have taken 1st RPIMa specialists as far afield as the Gulf, Bosnia and Africa.

The 1st RPIMa has several qualified IO—Intervention Offensive—combat swimmers, that were trained at Valbonne Diving School. They are equipped with Oxyger closed circuit respirators and various watercraft, including Zodiacs and sea kayaks. Counter-terrorist missions for the 1st RPIMa are more along the lines of close quarter battle and hostage release, as exemplified during Operation Amaryllis in Rwanda and Pélican in the Congo.

The 1st RPIMa is the largest of the COS units and it has a wide range of capabilities, making it well-suited for numerous types of special operations missions. Over the past fifteen years French commands have implicitly acknowledged the regiment's expertise by including elements of it in each major operation the Army has undertaken.

1st RPIMa WEAPONS AND EQUIPMENT

As a large unit with a wide variety of missions, the 1st RPIMa is equipped with an impressive array of different weapons and equipment. Even standard French armed forces weapons are adapted to fit the unit's specialized requirements for its reconnaissance, counter-terrorism and personal security missions.

Individual personal weapons include the Pamas pistol, the Smith and Wesson 686, the 5.56 mm Pamas with Trijicon ACOG scopes and the 9 mm Mini-Uzi with silencers for the combat swimmers (this weapon is a favorite of many special units because of its reliable functioning after prolonged immersion in water). As may be expected, the omnipresent HK MP-5s (A5, SD3 and SD6) are a part of the unit's arm-ory. These weapons have OWL sighting systems with infrared signature good to 25 meters. Personal security teams have MP-5Ks and Chief 60 revolvers, as well as the IMI Micro Uzi. Several of the smaller caliber weapons have been modified with either Aimpoint sights, the AIM laser designator or the Trijicon reflex sight.

The unit's armory boasts a complete range of assault rifles. It seems probable that the Famas is widely used for operations, but other rifles are widely used as well, such as the 5.56 mm M-16 A1 and A2, the M-16 A1 and A2/M-203, the M-16 A2 723 and 733 (with M-203), and the 7.62 mm HK G-3, with or without the 40 mm HK-79 grenade launcher. Incidentally, the *"bérets rouges"* of Bayonne, as the 1st RPIMa are often called, stand by the tried and true single round M-79. Although the AK-47 with foldable stock is not officially on the unit's equipment roster, it is nonetheless very present.

The unit is also equipped with a good variety of precision rifles, ranging from the 7.62 mm Giat FR-F2 to the HK G3 A3 with scope. For heavy sniping the 1st RPIMa uses 12.7 mm Barett M-82 A1s and the Hecate II. There is also the 7.62 mm Ultima Ratio Commando II heavy sniper rifle, made by PGM company and equipped with a silencer.

Assault teams are equipped with the paratroop version of the 5.56 mm Minimi light machinegun and the vehicle mounted AA-52 and heavy .50 caliber machineguns. Unit shotguns include the Mossberg pistol grip with Sure-fire tactical lights, as well as the Benelli M-3 with foldable stock and higher capacity (8 rounds) magazine.

Apart from the U.S. M-79, the single

At right:
The 1st RPIMa was the first military unit to be equipped with the integral silencer 7.62 Ultima Ratio heavy sniping rifle. This excellent weapon is capable of noiselessly eliminating targets during expatriate or hostage rescuing missions.

Below:
Following the 1996 Operation Almadin 2 in Bangui, 1st RPIMa prepare with General Thorette, French Armed Forces commander, to execute a "grappe" operation from an EOS Puma.

round HK- 69 A1, the South African rotating six round MGL, the vehicle mounted 40 mm HK GMG and the 52 mm individual Fly-K TN 8111 make up the unit's complement of grenade launchers. For mortars, the combat sections use the 60 mm TDA Comman-

do and the South African 60 mm Vektor M4 and M1 (with bipod).

The so-called "heavy" sections are equipped with 81 mm and 120 mm mortars, the Milan anti-tank weapon and LRAC, Apilas and Wasp 70 mm rocket launchers. They also have Stingers and Mistral air defense systems.

Night vision is facilitated by the OB-50 mounted on some weapons, and various night vision goggle and binoculars including the OB-42, the SFM "Clara", Agenieux's "Lucie" and the "Ugo", made by Thomson.

The 1st RPIMa has made remarkable progress in the field of communications gear acquisition. The unit now has VHF/FM TRC-5102 emitter/receivers, TR-PP 39 for intra-team communication, the long range TRC-350s and the TRC-743 and 745 tactical terminal equipment and the Immarsat image transmission kit. The unit's Airborne Commando Groups (GCP) are equipped with Questar 3 variable lens telescopes.

Below:
Sniping team armed with French weapons including a 7.62 Ultima Ratio with integral silencer and a 12.7 mm Hecate II. The Bayonne staff has been making a serious effort for some years to arm its troops with the best equipment.

The unit has been assigned light vehicles in order to perform long range motorized reconnaissance, including P-4s with mounted .50 caliber machineguns, recognizable by the cable cutting bar attached to the front, while awaiting delivery of the new light rapid reconnaissance vehicles (VRI—*Véhicules Rapide d'Investigation*). The unit uses a whole flotilla of ACMAT light vehicles, ranging from reconnaissance and communications trucks to wreckers. All-terrain vehicles are also used by teams in the field to link up with each other.

Water operations are carried out with semi-rigid Zodiac boats with 40 hp motors and sea kayaks.

1st RPIMa INSIGNIAS AND BADGES

Left to right and above to below:

—1st Airborne Infantry Company
—Current RPIMa Insignia
—Faculty: Overseas Airborne Brigade (1959-60)
—Instruction Group
—1st Company (Former)
—1st Company (Current)
—2nd Company (Former)
—2nd Company (Unit designed patch)
—2nd Company (Current)

—3rd Company
—4th Company
—Training Company
—Communications Company
—HAHO /HALO Team
—Long Range Reconnaissance Team (Airborne)
—LRP Team (Modified)
—RAPAS badge
—Band
—Commando Badge

—Military Assistance Mission (Chad)
—Manta DAO Detachment (Unit designed patch)
—1st RPIMa Sports Insignia
—GSA Insignia (Parent command of 1sr RPIMa)
—RAPAS badge (Dress)
—RAPAS badge (Fatigues patch)
—RAPAS badge (Modified)
—RAPAS badge (In silver, modified)
—Colonial Airborne Brigade (1955-58) and current 1st RPIMa unit shoulder patch

Preceding pages:
2 Company has several sections with the close quarters combat specialty, penetrating inside buildings, eliminating resistance and rescuing hostages.

At right:
1st RPIMa is currently making an impressive communications equipment upgrade, at the same time creating a transmissions specialty company. It has also made great strides in rapid and deep penetration reconnaissance, using VLRA light reconnaissance vehicles and the new soon to be delivered VRI rapid reconnaissance vehicle.

Following pages:
1st RPIMa has a wide range of arms at its disposal, including the Chief 60 revolver for bodyguards and the brand new Herstal P90 machine pistol. There is heavier stuff also, like the 120 mm mortar or the Mistral anti-tank missile, not to mention the « commando » selection of mortars like the 60 mm Vektor M 1.

FRENCH NAVAL COMMANDOS

(Les Commandos-Marines)

"From the time that COS was established, 18-20% of all operations carried out by Naval Commandos were for that command, except in 1997 when the figure was 61%", says Navy Captain Quentin.

THE BEST EXAMPLE of this is the *Commando de Montfort* ; who performed an offshore parachute drop by a French Navy ship, followed immediately by two exercises in Togo and the Ivory Coast. It was then deployed to Kinshasa for Operation *Pélican,* when in June, another overseas military operation for light units—called a DLMO (Détachement Léger Militaire Outremer)—came up. It was called Espadon, and involved an evacuation of expatriates from Sierra Leone. The group finally got back to its base at Lorient on 4 July with four missions under its belt.

As of 1 October 1993, COFUSCO got a "Special Forces use chart", drawn up by the staffs of the three services, putting 400 commandos at the disposal of COS. The chart specifies the capabilities of each Naval Commando unit capable of performing landings, assaults and reconnaissance either at platoon or company strength.

Staff at Lorient has three units at its permanent disposal, as well as one at Toulon and one at Djibouti. Stand-by equipment is stored at the three bases, and at Toulon gear can be loaded onboard a TCM or aircraft carrier. In 1997, during the Congo operations, the *de Montfort* units made a stop at Toulon before touching down in Gabon to link up with elements of the 1st RPIMa for the operation at Pointe-Noire.

A naval commando unit is made up of four 20-man platoons, broken down into two combat teams. There is a Headquarters section—ECT (*Elément de Commandement et de Transmissions)*—that ensures communications and command functions, an assault team, a reconnaissance team and a fire support team. Each element takes an alert watch, and a standard 48 hour alert is reduced to as little as 6 hours for these Special Forces units. As Lieutenant Coupanec, commander of the Trepel commando unit, explained in reference to the 1997 Congo expedition, "We had several sections at Toulon when the alert sounded at 2230 hours. COS gave us the word to move out to the Congo. At 0500 there were 25 troops on board

Preceding pages:
These Trepel commandos advance along a beach. The photo shows the different weapons used by this type of platoon. High volume of fire is stressed for these troops: Famas with OB-50 night scopes, the M-4 carbine with M-203 grenade launcher and the Minimi light machine gun. The unit also uses 12 gauge pump shotguns and the 60 mm commando mortar. Note the beret insignia for French Naval Commandos superimposed on photo.

At left:
This Trepel commando in night vision goggles and armed with an M-16/733 assault rifle with the M-203 grenade launcher, 150 round drum magazine and a visible/invisible laser designator provides excellent fire support to forward elements.

At left:
Two reconnaissance swimmers from the Trepel unit perform a beach check in waning light before the arrival of the other platoons. During a beach reconnaissance operation, beach slope is determined and, if necessary, a beach file is created. They may be required to hold the beach until the arrival of additional naval commandos. These men are armed with 5.56 mm Sig 551 assault rifles.

FRENCH NAVAL COMMANDO OPERATIONS
SINCE 1991 :

1991:
Evacuation of expatriates from Somalia.
Mine neutralization and embargo enforcement in Kuwait.

1992:
Operation *Hortensia*: beach and port reconnaissance.
Operation Iskoutir in Djibouti.
Bosnia: operations with the *Commando de Penfenteyo*.

1993:
Operations *Balbuzard* and Shape Guard in the Adriatic, lasting through 1996.
COS-directed Operation *Oryx* in Somalia, with the *Commando Jaubert*.

1994:
Expatriate evacuations from Yemen: *Commando de Montfort*.
Operation *Turquoise* in Rwanda: COS mission with the *Commando Trepel* providing security for civilians.

1995:
Operation *Azalée* in the Comoros with the *Commando Jaubert* and other Special Forces units.
Operation *Nautile* in the South Pacific (Mururoa). Security for testing bases provided by the Hubert unit.

1996:
Operation Badge in Afghanistan. *Commando Trepel*.
Operation *Salamandre* with IFOR in Bosnia.
Operation *Malebo* in Zaire: *Commando de Penfentenyo*.

1997:
Operation *Pélican* in the Congo: COS mission using de Montfort and Hubert units.
Operation *Espadon* in Sierra Leone with the *Commando de Montfort* .
Operation *Neptune* in the North Sea: *Commando de Penfentenyo*.
Operation *Alba* in Albania: beach reconnaissance and western expatriate evacuation by Hubert and Jaubert units.
Operation *Maracuja* in the Caribbean: *Commando Trepel*.
TAAF mission carried out by de Montfort and de Penfentenyo units.
SFOR missions in Bosnia with de Montfort and Hubert units.

1998:
Operation *Iroko* in Guinée-Bissau and *Malachite* in the Congo.

1999:
Operation KFOR in Macedonia and in Kosovo.

Above:
Naval commandos atop a VAB at the Brazzaville airport on 13 June 1997, during Operation Pélican 2. Several thousand expatriates were evacuated from Congo-Brazzaville during the operation.

At left, above:
A Trepel commando oversees a Hutu refugee evacuation in the mountains of Rwanda during Operation Turquoise in 1994. This was to be the third major COS operation for Naval Commandos.

At left, below:
Combat swimmers from the Commando Hubert performed flawlessly during their part of Operation Nautile when Greenpeace vessels entered French waters in 1995 at the nuclear test site of Mururoa in the South Pacific, in the face intense media coverage.

At right:
Naval Commandos from the Commando Jaubert on board a French vessel in the port of Düres, Albania in April 1997. These commandos had reconnoitered the landing sites prior to the arrival of French units.

Below:
De Montfort personnel prepare to carry out a reconnaissance mission in Brazzaville during Operation Pélican 2, undertaken to evacuate Western expatriates from Congo-Brazzaville.

Above, at right and following pages:
A series of photos showing drops or rappels from an 1st EOS Puma of the Special Operations Division. Each of the four assault sections has around twenty troops that do reconnaissance from helicopters; they drop or rappel from the helo into the water, recover their waterproof sacks and load them into a Zodiac and proceed to the beach to perform reconnaissance operations. Naturally, this is a night operation, originating far offshore out of sight of the objective. A Puma can drop off half a platoon (six men) in less than in a minute.

a C-130 with all their gear, headed for Libreville, in Gabon."

Less than 30 hours later, these troops were attached to a detachment of the 3rd RPIMa and some of them were being transported by helicopter to a Navy building rooftop facing Pointe-Noire. Meanwhile, other elements of the unit were carrying out an evacuation of expatriates with the 1st RPIMa at Dolissi, in the Congo. The ground troops were supported by an EOS Puma armed with 20 mm cannon and Mirage F-1s above.

The mission at Pointe-Noire was to provide security for the French Consulate and to round up expatriates scattered throughout the city. There were some hairy moments, especially when French units ran into Angolan troops

that had crossed into Congolese territory during combat with the government troops loyal to Congo President Lissouba.

The ECT sections were augmented from 8 to 20 personnel in 1997. The command cell is made up of the commander, his second in command, a Navy officer with an intelligence specialty, a VF communications specialist and Standard C transmission kit specialist—airborne qualified, but not necessarily holding the commando rating, two OPSON (*Opérateurs son*) transmitter operators, three ETRACO (*Engin de Transport Rapide de Commandos*—Commando Rapid Transport Aircraft) pilots and 10 enlisted men.

As its name aptly suggests, the assault platoon has the job of taking and hold-

ing a beach ahead of the main attack body. However, a very important peacetime function is to participate in Operation *Merlu*, an action aimed at enforcing French fishing laws on foreign fishing crews. The missions sometimes take on the aspect of genuine maritime counter-terrorism operations against armed and dangerous fishing crews engaging in illicit activities. These crews use everything they can get their hands on including steel bars, hooked gaffs and pump-action shotguns.

Teams of six to eight naval commandos patrol everywhere in the world where French waters can be found. As this was being written, men from the Jaubert unit were operating off the Kerguelen coast in Brittany, and in the French Antarctic territories.

The support platoon backs up the assault platoon with heavy fire support, including two 81 mm mortars, two Milan rocket launchers and 52 mm Fly-K grenade launchers. Also, over the last few years, there has been an intensive

effort by COFUSCO to provide several types of arms to shooters, depending on their mission type. Now each unit has G-3s, Ultima Ratios and McMillan sniper rifles, in addition to the FR-F2s. The heavy sniping rifle Gepard—14.5 mm—currently in testing, is also on the horizon. In general, commandos are armed with two types of personal weapons, including an assault or sniping rifle and a handgun.

About 75% of reconnaissance platoons are troops between the age of 20 and 25 and are commanded by a master chief petty officer with a petty officer as second in command, a petty officer third class and enlisted personnel. A reconnaissance platoon's mission is to reconnoiter beaches with the objective of determining whether it can be taken and held, with reconnaissance personnel responsible for holding the terrain until relieved. The platoon also compiles a beach feasibility study on the objective. Platoon members are equipped with Litton night vision goggles, marker buoys,

OPÉRATION "TARPON"

A *Tarpon* style operation consists of parachuting one or several platoons into the sea several thousand kilometers from France's shores. After recovering all their material, the commandos get on board a French Navy vessel that is patrolling in the area and proceed with a maritime operation, often a Zodiac raid.

Trials were made first with Super-Frelon helicopters and later, in 1985-86 with C-160s dropping personnel and containers of equipment, to test the feasibility of this type of operation. Six watertight containers containing a Zodiac and the equipment of three commandos were used for these tests. Although the equipment of an entire unit could be air-dropped, this type of operation would require very significant support. The *Tarpon* exercises take place several times per year, sometimes during the large-scale maneuvers in Africa.

soil sample equipment, infrared mark-
ers, underwater writing scrolls and AJM-
DLR laser designators as required.

Personal arms include the SiG 551
assault rifle and a PA Pamas—for beach
defense situations the commandos are
issued HK G-3s, Minimi light machine
guns, the FR-F2 and McMillan heavy
rifles.

Rear elements can be dropped or
can rappel from helicopters, parachute
into the sea, insert from submarines or
surface vessels or be winched down by
air using the "grappe" procedure from
helicopters.

Being dropped or rappelling from a
Puma or a Super-Frelon is very advan-
tageous. The drop-offs occur at 12 nautical miles from the objective and are
almost undetectable. Half a platoon
can off-load out of a Puma in less than
a minute, enabling a six-member reconnaissance element and with its gear in
waterproof sacks on board a Zodiac

ZODIAC HURRICANE PILOTS

The unit's Hurricane boats are genuine race boats drriveen by specially trained and qualified pilots and under the command of a team leader. The boats are powered by two 175 hp motors that take them to top speeds of 50 knots: boats like these require specially trained personnel to pilot them. To be admitted to training on this equipment, applicants must hold the commando specialty, have a boat pilot license and be qualified for rigid hull boat operation. The training lasts four weeks, the first of which covers such topics as all-weather navigation, water safety, nautical hazards, accessory equipment, towing, launching, piloting positions and docking. In the second week, high speed daytime maneuvers and maritime assault techniques are practiced.

The last two weeks are devoted to day and night high speed piloting, preparing and executing a long range raid and sea assault with two commandos on board. The qualification is obtained at the end of the training by passing an oral and written exam, and by the submission and approval by the commandant of the applicant's Hurricane pilot application.

Pilot training is constantly revalidated by supplemental coursework and missions. The commando team leader has training similar to that of watch chief, stressing the importance of precise navigation skills and a mastering of the instruments used in boat operations.

Every sea operation uses a minimum of two boats, with one serving as rally point for the first in case of trouble. Every Hurricane crew has night vision equipment and communications gear to keep in constant contact with command elements. The most sensitive moment for a high speed assault boat approaching a target is the wake crossing, which can easily capsize the fast closing Hurricane. The pilot must be adept with the rudder and the accelerator to successfully come alongside the boat to be boarded. This maneuver is repeated again and again during the qualification period, until it becomes second nature to the applicants.

Hurricane pilots successfully achieved the boarding mission on the Greenpeace vessel in Mururoa in 1995 with personnel from the *Commando Hubert*. They also work regularly with naval commandos during Atlantic and Mediterranean exercises, known respectively as *Armor* and *Estérel*.

with a 40 HP motor to accomplish a beach reconnaissance effectively.

Procedures for transporting Zodiacs in Pumas were set down in 1994-95 by EOS.

Navy Squadrons 32F and 33F can also handle this type of mission. The Super-Frelon has better transport capacity, with room and power for 12 men and 2 Zodiacs.

As Lt. Coupanec observes, "Recon platoon know-how is starting to spread among the unit's members. We now have 20 men capable of this type of exercise, but it hasn't been easy getting there. Also, it's important for expert recon members to maintain their skills."

As it stands, French Naval commandos are the most frequently deployed of the French Armed Forces special forces.

In fact, every operation involving French forces uses these "green berets", as they are called—*bérets verts*—proof of the recognition they have earned for their expertise.

Facing page, from left to right and top to bottom:
One of two Milan light anti-tank weapons organic to each of the fire support platoons of the various commando units. This type of weapon is essential for assaulting fortified positions or dealing with enemy armor. Unfortunately, it is impossible to use these weapons form light Zodiac boats.

All naval commandos are armed with the SiG-551 assault rifle with foldable stock, an excellent weapon that performs well under extreme conditions.

Assault platoons are equipped with the marine Remington 870 pump-action shotgun with foldable stock and extended magazine and high power light attached to the forward handgrip of the weapon.

Assault platoons also have HK MP-5 SD6 machine pistols with integrated silencers and laser sights.

Snipers are a big part of maritime counter terrorist operations. Commandos train regularly at sniping from EOS Pumas, here using a Famas with scope and straps across the door opening to help stabilize the weapon.

The LG1 Fly-K grenade launcher, shown here in firing position, is used in each platoon. This weapon is of large enough caliber and portable enough to be a truly effective support weapon.

Sniper firing from a Puma with a German 7.62 mm G3 SG-1 assault rifle and a Trijicon ACOG 3.5 x 35 scope.

French-made weapons are naturally a part of naval commando units' arsenals, especially the 7.62 mm FR-G2 sniper rifle. This aging weapon, much modified, is still an excellent rifle for medium range targets.

At right:
Heavy sniping is now done with the PGM 12.7 mm Hecate II, a weapon considered by specialists to be the best of its class.

Below:
The Hungarian-made Gepard heavy sniping rifle is a recent arrival to COFUSCO. The 14.5 mm rifle, currently being tested by Special Forces snipers, has impressed specialists, who consider it the best weapon in that caliber class. It can destroy a target from 2,000 meters.

INTERVENTION COMMANDOS MARINE

(Groupe de Combat en Milieu Clos

In 1994 COFUSCO set up a GIGN-inspired assessment process for candidates wishing to participate in a new maritime counter-terrorist organization. Since the operation was based on GIGN methods, it was clear that selective assessment of candidates would result in a "many are called, but few are chosen" situation.

O NLY THE BEST get through the selection process—about four per year. For one week, applicants—all experienced commando qualified personnel—are subjected to extreme conditions, day and night, including rucksack marches, physical training, face-offs with attack dogs to assess reactions, handgun qualification and other tests, all of which assess candidates' ability to react quickly and correctly.

There are 16 personnel in the GCMC under the command of an officer. The group is divided into two eight-man teams. Average age is 28 years: enlisted personnel are 24 to 25 years old, with team leaders and assistant squad leaders 34 and 30 years respectively. As needed the teams can be supplemented with maritime counter-terrorist (CTM) qualified personnel from other units. GCMC works regularly with B platoon of the *Commando Hubert*, a CTM specialized section of combat swimmers, during exercises at sea, both in the Mediterranean (the *Estérel* exercise) and in the Atlantic (the *Armor* exercise). GIGN personnel are often present during these exercises to harmonize procedures for joint CTM operations between the units.

The GCMC specialty is boarding a ship in terrorist hands using watercraft assault vessels, either Hurricane or Zodiac rigid hull craft. As one team mem-

ber explains, "Weather conditions often influence our approach: if the sea is flat or if there is fog, or if the ship is zigzagging, we use our fast attack boats because boarding by helicopter is too visible. On the other hand, in heavy seas helicopters are really the only way in."

Each of the 17 team members has his own specialty. This means that the unit has a qualified team member for most skill specialties, including weapons, munitions, demolitions, airborne operations, logistics, administrative, scaling, commissary, medical, communications, target analysis, diving, surface water operations, transport, movement in hostile territory, special assets (electronic surveillance, special explosives, etc.), intelligence and objective comprehensive reports.

Preceding pages, at left and following pages:
As of 1994, the French Navy Commandant can turn to two units for maritime counter-terrorism: B platoon of the Hubert unit or the GCMC close quarters combat team based in Lorient. This very specialized latter group has the mission of retaking vessels that have been overrun by terrorists. In the photo, an assault team makes its way along the passageways and the decks of a military vessel, after having fast-roped from a helicopter or boarded by grappling hook.

Preceding page:
The GCMC insignia appears superimposed on the photo.

MARITIME COUNTER-TERRORISM GEAR

Counter-terrorism equipment includes a jumpsuit and various accessory items. The suit itself is navy blue, identical to the one used by the GIGN, of thick, waterproof material. The body armor used is also a flotation device and the helmet and visor are bullet resistant. Other accessories include a blouse with multiple compartments, a gas mask and carrying case, a leather holster and a leather ammunition case.

GCMC WEAPONS

GCMC personnel from the two teams are equipped with S & W Stainless 686s, PA P-226s with laser beam and Sure-Fire tactical lights, MP-5s (SD3 and A5), G-3 assault rifles with Trijicon scopes, Famas and M-16s with 203s (40 mm rounds), naval version Remington pump action shotguns with the Sure-Fire tactical lights Minimi light machine guns and FR-F2 precision sniping rifles.

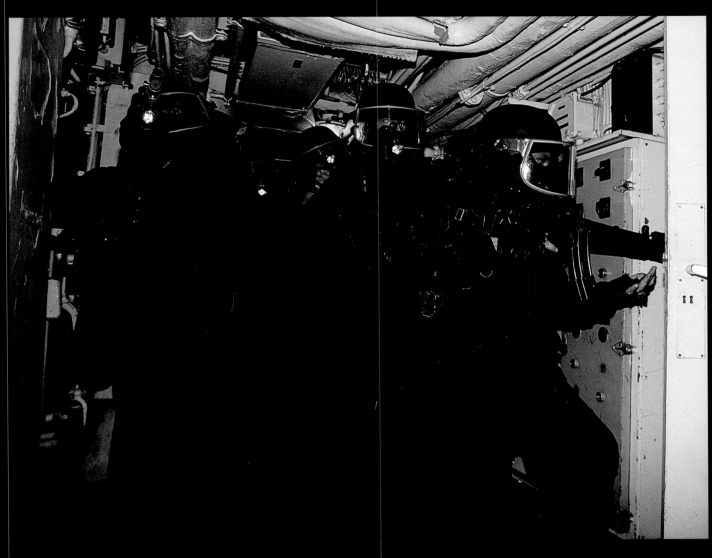

UNDERWATER ACTION UNIT

(COMMANDO HUBERT)

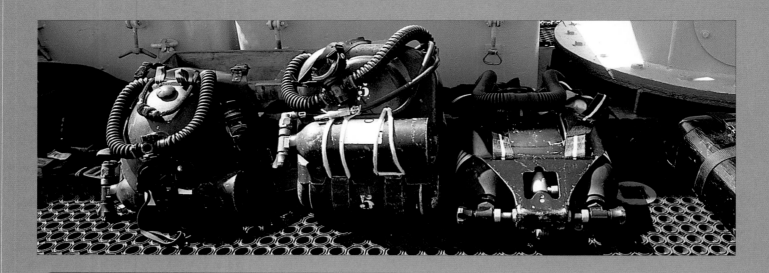

COMMANDO HUBERT EQUIPMENT

The *CASM* (*Commando d'action sous-marine*) underwater commandos are equipped with Oxygers respirators, for underwater demolition personnel, and with DC-55s and Oxymixgers for C platoon swimmer delivery vehicle operators.

The *Commando Hubert* has Vostock NG swimmer delivery vehicles capable of carrying two combat swimmers and 100 kilos of explosives. The vehicles are 6 meters long and 1 meter wide, they weigh 1,800 kilos and can move at 6 knots. They can be loaded aboard surface vessels but can also be strapped to the decks of Agosta class submarines in containers called "suitcases". The unit also has underwater traction gear.

Surface movements are accomplished in 7.33 meter Hurricane boats that weigh 2.5 tons and have an effective range of 100 nautical miles. The Hurricanes can carry 10 commandos, a pilot and a raid leader, and can withstand up to Force 5 seas. There are also 90 hp Type EF Zodiacs that are 5 meters long, have a 60 nautical mile range, can do 20 knots and carry four commandos and a pilot. These boats are good in up to Force 3 winds. Futuras have 40 hp motors, are 5 meters long, can go 60 nautical miles, have a maximum speed of 12 knots and can withstand up to Force 3 winds. They weigh 350 kg and hold six commandos and a pilot. The Nautiraid sea kayaks that travel at two knots and

The *Commando Hubert*, an Underwater Action Commando Unit (CASM—*Commando d'Action Sous-marine*) is one of the naval commando units that make up COFUSCO. The unit is based on the Saint Mandrier peninsula, just opposite Toulon.

This unit of combat swimmers was formed in Algeria, but in the fifties chose Toulon as its base. There are several reasons behind this choice of location. The military diving school is close by and, unlike the Atlantic, the Mediterranean sea has no daily tide, meaning that training is not limited to certain hours. The location also suits the requirements of French defense policy because it is an excellent take-off point from which to launch the unit rapidly on special missions, either for the French Navy or for COS.

The minimum qualification for becoming a member of this combat swimmers unit is to have served four years in a Navy unit and to have a diver's rating. An application for admission to the unit is submitted, and if accepted, the candidate attends an assessment course known as the Combat Swimmer's Course (CNC—*Cours de Nageur de Combat*) at the Saint Mandrier Diving School. A handful of successful candidates receive their combat swimmer qualification at the end of this process and are assigned to the Hubert unit for an initial period of three years. In reality, they never spend their entire tour at the unit's headquarters, because they must obtain the advanced qualification in their specialty, which implies a six month stay at Lorient, as well as the airborne qualification leading to master parachutist or trainer/drop specialist, meaning another five or six months at Pau.

The "new" combat swimmer is usually extended with the unit for another three years after this period, where he can either become a dive instructor or be sent to Lorient to improve his capabilities. The average age of permanent duty personnel at Saint Mandrier is thirty, they all have at least a high school education and 70% of them are mid-level NCOs.

The Hubert unit is divided into two companies, the 1st, which is an operational unit, and the 2nd, which has a combat support role. In addition, the unit has a Navy vessel attached to it, *Poséidon*, that serves as a surface base for combat swimmers and two-man swimmer delivery vehicles (called PSM—*Propulseur sous-marin*).

The 1st company is made up of fifty

combat swimmers, divided into four platoons, A, B, C and D. A platoon is the command and control section, and is equipped with Hurricane rigid watercraft. B platoon is a maritime counter terrorism team (CTM—*Contre-terrorisme maritime*), while C platoon specializes in underwater operations. D platoon handles reconnaissance and support operations. When operations are being planned, the unit commander can pan through the different platoons to select personnel among the combat swimmers with the qualifications he needs for a given mission. Lt. Commander Rebour, chief of the *Commando Hubert*, states, "Each swimmer is a unique individual with a different combination of skills. My job is to marry up the skill to the mission as effectively as possible. To do this I have available a pool of talent containing the most experienced members of the unit including a maritime operations second in command, a PSM group, an operations bureau and armament and munitions sections."

B platoon's CTM specialty is the equivalent of the GCMC Close quarters combat group based at Lorient, with the particularity that it is defined as a unit "with

underwater operations capabilities". This makes the unit the most specialized counter terrorism unit with a naval alignment for assisting the land oriented GIGN in operations. The two units work together closely during counter-terrorist exercises, with the Hubert elements acting as "pilots" during water environment parts of the exercises.

C platoon is comprised of Swimmer Delivery Vehicles (PSM—*Propulseurs Sous-Marin*) and their crews and maintenance teams for the equipment. Its mission is to get combat swimmer teams near the objective, setting off from a surface vessel or a submarine. D platoon's reconnaissance and support specialty

leaves it no time for either the PSMs or counter-terrorist actions. These men are specialists in firepower and explosives well as heavy sniping with 12.7 mm rifles. They are HAHO/HALO qualified and perform underwater demolition operations. D platoon is combat swimmer qualified and do channel and beach reconnaissance as well as development and testing of new equipment.

Future plans call for the creation of an E platoon that will have crews of rigid and semi-rigid water craft; equipment will be taken from D platoon and personnel from 2nd Company.

Hubert unit draw from 2nd company, a support unit made up of 30 men. The company has several teams that have such specialties as communications, swimmer delivery vehicle maintenance and repair, commissary, maintenance and boat operation. The *Poséidon* serves as a support vessel during training operations. The 200 ton, 40 meter vessel has a crew of 14, can make 12 knots and has a decompression chamber.

Although the *Commando Hubert* is a COFUSCO unit it is completely unique, with underwater capabilities specific to no other organization of the French Navy.

(Commando Parachutiste de l'Air n°10)

COMMANDO PARACHUTISTE AIR

10

Two preceding pages.
CPA 10 teams reconnoiter areas for assault landings and bring airport facilities back on line after attacks. This doesn't mean that they can't handle laser designation missions. To accomplish these, the unit has a laser guidance team in each of its cells. These teams have had the French-made laser designation equipment Cilas DHY-307 since 1997. Note CPA 10 field unit patch superimposed on photos

Opposite and below. Despite its reduced size, CPA 10 unit members each usually have several qualifications, making them sought after for numerous missions, especially airport facility rehabilitation. This kind of operational capacity is essential for airlanding units in hostile territory.

In 1994, Air Force Commando Unit 10 (*Commando parachutiste de l'air N°10—* French Air Force Commando Unit 10, or CPA 10), in conjunction with the Special Helicopter Operations Division (*Division des opérations spéciales hélicoptères—DOS/H*) took on the responsibility for commando operations for the French Air Force.

AS OF THAT TIME, CPA 10 fell under the aegis of the COS *(Commandement des opérations spéciales)*, which is the equivalent of the American USSOCOM. COS acts as the go-between for the commando units and the Commander in chief of Armed Forces (CEMA—*Chef d'état-major des armées*), for whom it directly executes operational requirements. This operational designation falls under the responsibility of the Special Airborne Infantry Command (CFCA-*Commandement des fusiliers de l'air*).

The soldiers of this elite unit come mostly from CPA 40, the component with which CPA 10 forms the French Air Force commando unit (EICA--*Escadron d'Intervention des Commandos de l'air*), based at Orléans (formerly at Apt, France). Applicants to the unit must be non-commissioned officers and professional military. A typical individual profile for this unit is as follows: 27-28 years old, minimum rank of sergeant, Airborne qualified, level II commando qualified and five years time in service. Officer and enlisted applicants to the unit must go through a nationwide selection process. Upon successful completion of this process, new CPA 10 soldiers are divided into teams for specialized training in such fields as commando trainer, combat parachutist, sniper, combat search and rescue specialist and other fields.

HEIRS OF FRENCH AIR FORCE INFANTRY TRADITIONS

CPA 10 is one of the units descending from the original Air Infantry Groups, including the 610th GIA based in Reims, France and the 602nd in Baraki, Algeria, both activated in 1937. In 1943, GIA airborne troops from these outfits combined to form the 1st Airborne Infantry Regiment (1st RCP—*1er régiment de chasseurs parachutistes*), which took part in campaigns in Eastern France in 1944 and 1945. On 1 August 1945, the 1st RCP was transferred to the French Army, leaving the French Air Force without airborne assets. In 1956, French Air Force Commando units were re-established. The units formed were the 10th, 20th and 30th Airborne Commandos. The following year saw two other units, the 40th and 50th added to the group. But by 1961, only the 50th Airborne Commandos remained operational.

In 1965, a squadron of Air Force commandos was established in Nîmes, France, with the mission of training Special Forces troops in air base perimeter defense. In seven years the number of new units emerging in this field justified the creation of an Air Force Commando Unit. Initially, the group was called EEI—*Escadron d'Evaluation et Intervention*, and had a mission of reconnoitering and direct action in 1978. However, by 1992 the squadron's role had changed to one of personal security and direct action in conflicts. Its name was changed to EPI— Escadron de Protection et Intervention—and the group became one of the top seven elite units in the French Armed Forces. Two years later, in 1994, EPI was designated an EICA, or airborne combat commando squadron, and split into two units, CPA 40 and CPA 10; the latter was given the Special Operations moniker. Personnel from these two units have taken part in operations in Mauritania, Lebanon, Chad, Saudi Arabia, Rwanda, Zaire, the Congo, Egypt, Oman, Kuwait and the former Yugoslavia.

Above.
The 602nd GIA (Airborne Infantry Group) pennant. This unit is at the root of CPA 10's origins.

Preceding pages.
A direct action team following a grappe and rappelling exercise. The photo was taken in the Apt region, where the unit is based until it takes up residence in Orléans. The unit is made up of three action teams, a command section, a training section and a logistics arm.

Opposite. **CPA 10 became one of the group of elite COS units in 1994. Since then it has been tasked with French Air Force special operations along with EHS in Cazaux and the DOS C-160 and C-130 squadrons in Toulouse and Orléans. While CPA 10 is only attached to COS for use in missions, it does the majority of its operations for that command.**

ASSESSMENT

Assessment for induction into CPA 10 occurs over a weeklong period with criteria for selection provided by GSIGN (*Groupe Spécial d'Intervention de la Gendarmerie Nationale*—Special Action Group) elite elements of the Gendarmerie (French military police force). This phase assesses physical fitness, airborne qualification test, timed swim tests, underwater exercises with wrists and legs bound, timed combat and "challenger" obstacle courses, military orienteering exercises, endurance exercises, compass exercise, sleep deprived weapons firing situations, motivation and stress management assessment.

A second phase assesses technical capabilities including familiarity with equipment, Special Forces and airborne techniques. In this phase instructors test candidates' tactical knowledge, ability to plan field maneuvers and to carry out a combat team scenario. The applicants are given a battery of questions concerning general knowledge, current events, history and geography. The selection process culminates with an appearance by each candidate before a selection board that sounds out his reasons for wanting to join CPA 10. The board's comments are considered jointly with those obtained from an individual interview with a psychologist. A final determination concerning a candidate's aptitude is made by EICA commanders that review each applicant individually.

Opposite and below.
CPA 10 has a HAHO/HALO team equipped and trained to transmit essential intelligence from long range destinations. The team is made up of an advanced controller officer, an NCO assistant team commander, a communications specialist, a radio operator, a medic and three riflemen to provide security.

CPA 10 is a company size unit of 120 men (6 officers, 35 non-commissioned officers with the rest divided into senior corporals and various airborne and infantry technical specialties).

The unit is organized into a command cell, three "action", or combat detachments, a specialized training cell and a logistics team. All three of the action detachments are combat teams with similar missions, but only one has HALO/HAHO capability and is made up of master parachutists that operate well within enemy lines. Several soldiers within this detachment have the tandem qualification and are consequently capable of placing non-airborne qualified technical personnel into position for restoring mission capability to strategic facilities.

CPA 10 communications assets include a transmissions team operating in the command section and under the direct authority of the team leader.

This team maintains radio communications via satellite using the Immarsat communications kit between forward operations and command centers. The unit is also equipped with photographic and video equipment to provide real time intelligence to headquarters.

COS designated missions for CPA 10 include laser target designation and laser target ranging. Each combat detachment has elements specializing in reconnaissance, securing and marking landing zones for air assault or airdrop operations and securing and restoring the operational capacity of airport facilities .

These units also provide typical special operations support in the form of infiltration, combat and exfiltration.

There are several teams within CPA 10 that handle target designation assignments. According to Major Phillipe Landicheff, commander of CPA 10, "The conflict in former Yugoslavia involving the insertion of forward controllers in 1994 reminded many of operations using tactical guidance personnel during the Cold War. Initially, French commanders in Bosnia deployed a forward controller and a pilot in a VAB (*Véhicule de l'Avant Blindé*—Front Armored Vehicle) to guide air elements to their targets."

During the Gulf war, airborne designators were used for illuminating targets.

Laser interference problems caused by natural obstacles or the weather, including clouds, were encountered during this period.

In 1994, COS acquired the U.S. built LTM-91, then in 1997, the French DHY-307 made by Cilas. The equipment's range is 5km, though according to most users, "...the closer to the target, the better the chances of maximum reflection". A forward controller (an officer) leads a laser designation team, with an NCO assis-

At left, below and on opposite page. **CPA 10 teams train for RESCO (Combat Search and Rescue) operations both for the French Air Force and for COS. This type of mission is very difficult and involves several helicopters and teams of commandos. It is an extremely complicated and risky operation.**

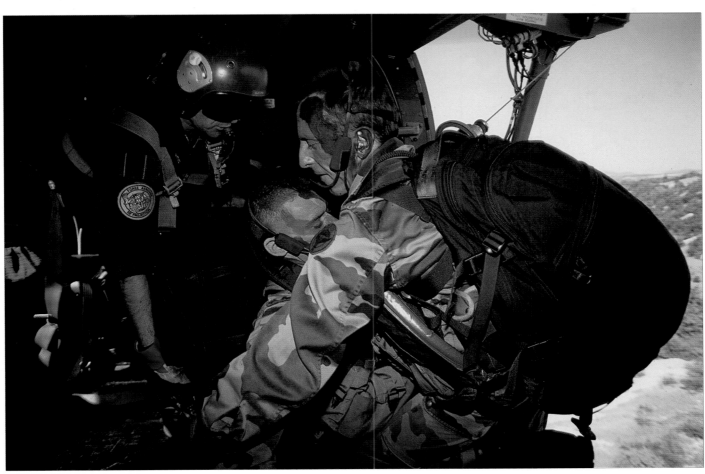

tant commander, a communications specialist, an operator, a medic and a three-man security detail.

CPA 10 personnel are considered the COS specialists in securing and restoring operational capacity to airport installations. Captain Francis Noel, the unit's second in command, says," We have to be ready, any time night or day, to quickly assess the condition of airport facilities and to make them operational in the shortest time possible. The unit has the radar, communications and cipher specialists, the electricians, armorers and electromechanical personnel required to repair most types of equipment damaged during an airport assault."

Another CPA 10 mission, known as RTPA in French—Reconnaissance de terrain de pose d'assaut—is to reconnoiter potential landing zones to identify, mark and prepare them. RTPA teams are equipped to rapidly assess, with the help of special soils bearing test equipment, whether or not a certain zone can support tactical aviation equipment.

The unit also receives training in Combat Search and Rescue operations (called RESCO—*Recherche et Sauvetage au combat*) for recovering downed personnel behind enemy lines.

This is an extremely difficult and dangerous mission, carried out by several teams, helicopters and support and surveillance assets. There is absolutely no room for error in this type of operation.

Before COS was established, the French Air Force had no special operations units. Now it has a superior asset, capable of air operations, reconnoitering and securing landing zones and airport facilities restoration that no other French armed forces component has the capacity to undertake.

*Preceding page, at left and above, following pages.
CPA 10 missions include, apart from reconnaissance
and target designation, the restoration of airport facilities to operational capacity to receive cargo aircraft.
CPA 10 is the only unit capable of this type of
operation, boasting several commandos trained as
radar operators, electronics and communications specialists and electricians. This series of photos show a
unit's progression through airport facilities, initial
reconnaissance to on-site repairs of damaged
facilities.*

MAJOR CPA 10 OPERATIONS

· **September 1993:** first deployment of forward laser designation teams in Bosnia.

· **June to August 1994:** elements of CPA 10 take part in the inter-army Operation Turquoise in Rwanda and in the setting up of a forward operations base at Kisangani in Zaire (now the Republic of Congo).

· **May 1994 to 1997:** assignment of an operational detachment as a part of the multi-national Operation Crécerelle, then in Operation Salamandre at the U.S. military base at Brindisi-San Vito, Italy.

The detachment trained with members of U.S. CSAR (Combat Search and Rescue) teams that specialize in extracting aircraft crews stranded in hostile territory. Participation in these types of missions by various detachments from CPA 10 and CPA 40 during this period enabled personnel to perfect Search and Rescue techniques.

· **June-July 1995 :** during this period, elements of CPA 10 took part in Operation Balbuzard in Bosnia

· **August-September 1995:** during Operation Salamandre the team stationed at Birisindi, Italy carried out a search mission in conjunction with U.S. special forces personnel in an attempt to rescue the crew of a French Mirage 2000 downed in Serb-held territory in Bosnia.

· **December 1995 to October 1996 :** A liaison team was assigned to the French detachment deployed in Bosnia to support IFOR operations.

· **March 1997:** CPA 10 teams took part in Operation Pélican involving the evacuation of French nationals, then carried out target designation missions in support of combat aircraft during Operation Pélican 2.

· **January 1997:** As a part of SFOR in Bosnia, a CPA 10 team completed liaison and PSYOPS missions

under the aegis of COS.

· **March 1997:** The RESCO (French version of CSAR) team based at Brindisi took part in Operation Alba in Albania, with the objective of evacuating European nationals assembled on the beach at Dorves.

This operation was complicated by locals who threatened to storm the helicopters in an attempt to flee the country.

· **July 1997:** A liaison team was sent to Albania to support Operation Alba.

In addition to these operations CPA 10 teams took part in other missions and exercises abroad carried out by COS in Oman ; Jordan, Egypt, Great Britain, Senegal, Ivory Coast and Tunisia in conjunction with special forces personnel of those countries.

Opposite page and above
From left to right, view of CPA 10 commandos specialized in taking strategic airport buildings, especially the control tower. They are armed with the Famas assault rifle, the short-barreled Mossberg pump-action shotgun with pistol grip and Mag-lite attached to barrel, an HK MP-5 SD6 and Remington pump shotguns with foldable stock and eight round magazine.

Above.
CPA 10 uses Inmarsat Standart C and Merit communications kits (shown here) to transmit images to command centers.

Below.
CPA 10 uses 60 mm Fly-K individual grenade launchers and multiple grenade launchers mounted of P-4 light vehicles.

Opposite.
Sniper teams have FR-G2 heavy sniping rifles. For long range sniping they are equipped with 12.7 mm McMillan heavy sniping rifles. This weapon is especially effective for stopping counterattacks launched against units holding airports until the arrival of reinforcements.

Opposite.
CPA 10 commandos use 60 mm TDA MO mortars in addition to the Fly-K individual grenade launchers.

Opposite.
The Air Force commandos were among the first French forces to receive Famas assault rifles with 40 mm grenade launchers attached. These launchers are called M-203 PI and are very accurate when used against enemy positions.

Below.
For airport assault operations commandos often call on the 58 mm Wasp rocket launcher.

CPA 10 ARMAMENT AND EQUIPMENT

Unit weapons include the Pa Pamas G1, the Famas assault rifle, Famas with M-203mm PI grenade launcher, the HK MP 5 SD6 light machine gun with LTM-91 laser designator, the DHY-367 and AIM-1 laser designators, the Minimi light machine gun, the FR-G2 and McMillan sniper rifles, the 58 mm WASP rocket launcher, and individual and Fly-K multi-tube grenade launchers.

CPA 10 Communications equipment includes the Motorola Saber, the PR4-G, the TRC-350, the Thermit cipher unit, the Immarsat Standart C satellite relay kit and the Merit image transmitter.

ALAT SPECIAL OPERATIONS DETACHMENT

(Détachement des Opérations Spéciales de l'ALAT)

Above:
These Pumas are being reassembled immediately
after being unloaded from cargo aircraft on 2 October 1995
in Mayotte during Operation Azalée 2.
1st RPIMa and GIGN teams were transported by Puma
to their airport and the beach in front of the French embassy
objectives during this operation in the Comores archipelago.

Below:
This EOS Puma carrying General Thorette,
the operation commander accompanied by a security detail,
has just landed, during the May, 1995 Operation Almandin 2
in Bangui (Central African Republic).
DAOS helicopters were also used to transport rebel chiefs
during negotiations.

116

Preceding pages:
Pumas in flight as seen with night vision goggles. Regular training ensures that EOS crews are qualified in low altitude night operations, shipboard landings at day and night, parachute operations, grappe operations and naval commando sea drops.

Above:
ALAT detachments attached to special operations units have been equipped with ten different types of Gazelles. The greater number of aircraft allows these units to take on more missions like providing security for airmobile commandos in Pumas.

The ALAT Special Operations Detachment (DAOS), based at Pau-Sauvagon, is made up of ALAT(*Aviation Légère de l'Armée de Terre*—French Army Light Aviation) helicopters assigned to Special Operations. This recently organized unit supports all Special Forces by ensuring air transportation during day or night tactical penetrations, and other aviation support.

T HE 1ST SPECIAL OPERATIONS Squadron—1st EOS—is the transportation arm of DAOS: the firepower is provided by 2nd EOS, the other half of DAOS. 1st EOS has 12 helicopters, Pumas and

Cougars, originating from ALAT stocks that were "mothballed" after the major reorganization that occurred in the French Armed Forces.

Both Cougars and Pumas are equipped for night navigation and can be armed to fit the mission, with 20 mm side mounted cannon or 12.7 mm machine guns.

Several of the helicopters have the "Chlio" thermal cameras, rendering them CSAR capable. Crews specialize in night flying, which accounts for one third of their missions.

The 1st EOS squadron supports COS units during fast-roping, rappelling, grappe (helicopter insertions and extractions of troops that dangle from a cable like a bunch of grapes); night LZ touchdowns, heavy firepower support, maritime insertions of

117

At right above and opposite: **Several views of one of the EOS Pumas on the tarmac at M'Poko Airport in Bangui. This one, with 20 mm door cannon, took part in the successful Almadin 2 action in support of government troops against rebels at the beginning of January 1997.**

At right:
Shoulder patch of the 1st Special Operations Squadron of Pau, that have Pumas and Cougars.

personnel and small equipment and shipboard counter terrorist assaults.

The 2nd EOS uses 10 armed Gazelles to provide Special Operations units daylight and night time air support.

The Gazelles can provide units on the ground with very significant firepower in terms of vol-ume and accuracy, thanks to the squadron's exceptionally wide range of equipment, including the Gazelle-Viviane, the Gazelle with side mount-ed 20 mm cannon, the Gazelle-Mistral and two specially equipped command-and-con-trol helicopters. The Gazelles can provide in-flight security for the transport Pumas in addition to their other fire support duties. It is possible to load two Gazelles into a Transall, making helicopter provided transport and fire support available rapidly to deployed COS detachments.

There was a selection process for DAOS crews in 1998, open to all ALAT pilots and mechanics on a volunteer basis. The procedure was made very difficult, due to the priority given to maintaining a Special Forces esprit de corps among successful can-didates.

Above and Below.
Views taken from behind the cockpits of a Gazelle (above) and a Puma (below) in a night vision setting.
As units dedicated to Special Operations, these two EOS squadrons train mostly at night for Special Forces missions carried out by elite unit teams.

Opposite
Now that 2nd EOS Squadron has Gazelles with 20 mm cannon the unit can provide far superior support fire for ground units in action situations. Gazelles are also capable of providing security to the Pumas during commando actions, unlike Pumas, even those with 20 mm cannon. In addition, Gazelles can perform night missions, like the one shown here using night vision equipment.

Opposite.
Having Gazelles as assets is a clear advantage for COS. The command can field two of these helicopters very rapidly by Transall, one with 20 mm cannon for heavy fire support and another capable of transporting two or three troops to an objective.

Below.
Trepel naval unit commandos from a reconnaissance platoon drop into the sea off Lorient with their equipment in waterproof sacks and their Zodiac, headed for a beach reconnaissance mission.

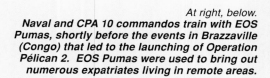

At right, below.
Naval and CPA 10 commandos train with EOS Pumas, shortly before the events in Brazzaville (Congo) that led to the launching of Operation Pélican 2. EOS Pumas were used to bring out numerous expatriates living in remote areas.

SPECIAL OPERATIONS DIVISION (Division des Opérations Spéciale

The COS has a certain number of highly skilled air crews teams at its disposal to assist in special air operations. These crews are capable of accomplishing difficult missions carried out at night and in hostile territory. The crews are part of the DOS (*Division des opérations spéciales*—Special Operations Division) and are attached to a unit called the CIET (*Centre d'instruction des équipages de transports*-Air Transport Crew Training Center) based in Toulouse, France.

THIS GROUP HAS ONLY Transall aircraft. If other equipment is required, COS can turn to C-130 Hercules crews based in Orléans, where CIET has a unit to administer these assets.

The DOS unit in Toulouse is commanded by Major de Rancourt, with Captain Xavier Pexeddu second-in-command. Personnel is organized into 17 crews, containing 16 naviga-

tors, 6 pilots, 4 weapons systems officers, 6 flight mechanics and 1 ground mechanic.

While the DOS does not have aircraft included in their assets, CIET has a standing order to make one C-160 R (Modified) available to it every day for training, for large-scale foreign exercises and when an operation is launched.

Personnel assigned to COS, all originating from Airlift squadrons, are officially part of FAP (*Force Aérienne de Projection*—Forward Air Group). As such, they are still required to perform FAP missions, mainly for CIET. DOS air crews are the most experienced in the entire French Air Force and are therefore the most logical choice for instructors at the training center.

Despite this activity, COS missions remain a priority, as is readily discernable after a glance at crew flight missions : 70% are DOS operations, with 20% CIET training and 10% FAP missions.

To get assigned to DOS, personnel must be professional military, be

Preceding pages:
View of a DOS cockpit before a night flight. Pilots use OB 56 night vision goggles and C-160 instrument panels are specially treated to enable safe operation of equipment during low level night flying. Note unit in-signia at lower right of page at left.

Below:
Two pilots and a navigator prepare for a night flight at their Toulouse base. The navigator, at center, enters data into a computer which will then be transferred to the Transall flight computer.

At right, above:
One of the two pilots shown prepares to embark on a night flying exercise with OB-56 second generation night vision goggles. Each DOS crew executes at least five low-level approach night landings with night vision goggles.

appropriately qualified pilots or mechanics and have a minimum of between 1,800 and 2,000 flight hours. The most experienced of the flight mechanics have up to 6,000 flight hours. Captain Puxeddu affirms, "Although personnel coming into DOS have all the basic qualifications, all of them need six months additional training to acquire techniques used in this unit. Only personnel holding the various NVG (Night Vision Goggles) qualifications are assigned to the detachment".

A DOS crew is made up of two pilots, a NOSA (*Navigateur et Officer Système d'Armes*—Navigator and Weapons Systems Officer) navigator and two mechanics (a flight and a ground mechanic). The average age for pilots is 30 and 35 for mechanics. While every operation is different, it generally requires three hours for a crew to prepare before setting out on a mission.

If the mission requires, crews can be transported to a friendly air base where it can secure a Transall and begin the operation. Travel is usually accomplished on board a French Air Force DC-8 or Airbus in the company of an elite unit.

DOS missions are very specific and are carried out in support of elite units like CPA 10, 1st RPIMa and the Naval Commando units. They are generally night missions involving infiltrations and exfiltrations in support of equipment drops or airborne landings, as well as day and night low altitude airdrops.

They also include day or night high altitude jumps to 7,000 meters, and can involve personnel with or without oxygen performing HAHO jumps or vehicle drops.

Another DOS mission is forward area helicopter refueling. DOS, in conjunction with French Air Force personnel designed a specialized refueling vehicle that they call the FTM

(*Fardier Technique Modulaire*). This vehicle is capable of refueling two Puma helicopters in thirty minutes, from the moment the Transall touches down to its takeoff.

Yet another of the unit's capabilities is in radio relay missions. DOS

Below:
The Fardier helicopter refueling vehicle, designed by DOS for refueling from a Transall.

DOS was established in September 1993 at Orléans, and was made up of Transall crews. The unit, commanded by Major Bernard Hufschmidt, was originally a part of the 3/61st Airlift squadron stationed in the Poitou region of France. At this time crews began NVG night flying training.

Between 19 June and 11 August 1994, a DOS Transall was sent to Zaire for operation Turquoise to perform supply runs between the airports of Goma, with a long airstrip for heavy cargo traffic, and COS units at Bukavu. More COS Transall crews came in to relieve and supplement the first crew, carrying out shuttles to resupply COS helicopter operations in Rwanda and a naval commando airdrop at Butare. Later, a Transall conducted a night airborne landing on a zone reconnoitered by special forces. The flight originated in Toulouse, France and passed through Heraklion (Greece), Djibouti, then Mahé (Seychelles) and Djaouzi in the Indian Ocean.

In 1997, DOS Transalls participated in various operations and exercises including a French-Togolese joint mission, operation Pélican in Libreville between 24 March and 10 May, a Jordanian operation in October and a training mission in Scotland in November.

aircraft are equipped with radio and information linking systems that enable communication between COS units on the ground and command staff in the air or in Paris. This amounts to an air-mobile command center.

DOS Transalls can also support PSY-OPS sorties by dropping tracts, perform sea drops of rigid or semi-rigid vessels for naval commandos and provide logistical support for COS units.

DOS crews demand a lot from their Transalls: night flying, serving as flying command posts, refueling helicopters and playing the role of all terrain cargo transport system.

In fact, they are completely self sustaining on ground. As a part of an elite group, they go into deserted operations areas alone, or with support detachments, and must provide for themselves. Containers are stocked in the aircraft with enough supplies to sustain crews for several days in order to remain in close contact with COS units on a 24 hour basis.

Considering that DOS equipment is in most cases no different from what conventional units have, it is remarkable that the units can accomplish what they do.

But then, the unit does have among the best air cargo pilots and crews in the French Air Force, who regularly engage in night training operations involving low altitude flight and landings on very short runways.

Above:
The CIET unit insignia (Airlift Crew Training Center)

Below:
DOS Transall executes an assault landing in the Jordanian desert at Al Hafira in October 1997. DOS aircraft performed numerous night takeoff and landings during these exercises.

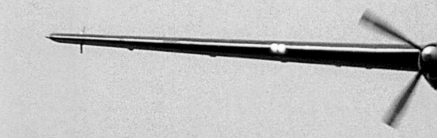

At bottom:
One of the first live exercises involving refueling of an EOS Puma from a Transall in the Jordanian desert, in 1996. Using the FTM refueling vehicle, developed by DOS and French Air Force mechanics, two Pumas can be refueled at night with rotors turning.

DOS CREW MISSIONS

Between September 1997 and September 1998, a typical DOS monthly training schedule (based on 26 working days) included 9 days of COS missions, 3 days of CIET missions, 3 days of CFAP missions, 1 day on miscellaneous maintenance or training functions, 3 days off and 7 all-night missions.

On an average, DOS crews can expect from 80 to 120 all-night missions per year. Crews spend 50 % of their time in the air, 80 % of which are night NVG flights.

A DOS crew effects one COS mission per month, one low altitude (100 meters) daytime mission and at least 5 night low level NVG landings.

FLYING GEAR

During combat missions each DOS crew member is equipped with a flight suit, a flight jacket, a flak jacket with survival gear, a PA MAC G-1, an Aerazur parachute, and second generation OB-56 night vision goggles.

Navigators and hold technicians are equipped with second generation IL Lucie Angenieux night vision goggles.

Above:
The navigator's mission during a night approach for an assault landing is to assist the pilots visually, using Lucie night vision goggles.

Zaire, August 1994. This DOS Transall departs from Goma on a re-supply mission for COS troops deployed in Rwanda. Transalls are lighter and have 8 wheels, making it possible for them to land on short runways, unlike the much heavier Hercules with only 4 wheels.

HELICOPTER REFUELING

Only DOS is capable of helicopter refueling operations, a mission it accomplishes using the Transall. From the very beginning of Special Forces operations in France, COS commanders understood that helicopter assets could only be re-supplied from secure, rear area bases that were generally very distant. This affected units' capabilities of ensuring readily available, mission-capable rotary winged aircraft for spot missions. To resolve this dilemma, French Air Force mechanics and personnel designed a lightweight vehicle that could be airlifted to refuel helicopters using fuel directly from Transall tanks. After several trials, Auverland, a military vehicle firm, was selected to provide its A3 model to which French Air Force workshops fitted—at lower than bid price—an excellent fuelling mechanism, particularly well suited to the mission, that was christened *Fardier Technique Modulaire* (FTM).

This vehicle needs only four minutes to set up once the Transall is on the ground. From that time, it takes only thir-

ty minutes to completely refuel two Pumas. The only other armed services in the world with this type of mission capacity are the British, the Canadians and the Americans, who use trailers. The Americans need an hour and a half and ten people to do the job, using a dedicated Special Forces aircraft. DOS only uses six people, three of which are Transall crew members.

In 1997 DOS personnel conducted exercises in Jordan with special forces from that country that enabled it to perfect the technique for refueling Puma rotary wing aircraft. Now DOS crews are capable of performing night NVG re-fueling missions.

FTM SPECIFICATIONS:

Length: 4.60 meters, **Width:** 1.70 meters, **Height:** 1.67 meters, **Ground clearance:** 22 centimeters, **Weight:** 1,800 kg

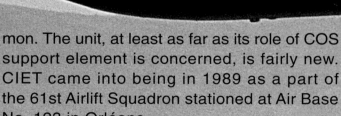

The official name for CIET Station no. 01.340 is the Airlift Crew Training Center. It is located in Orléans and is, in a way, the C-130 equivalent to Special Operations Division (DOS) based in Toulouse, which is equipped with Transall C-160Rs.

DESPITE THE OFFICIAL DESIGNATION, all COS personnel refer to the unit as DOS/C-130 and have come to recognize it through its participation in different exercises they have conducted in com-

mon. The unit, at least as far as its role of COS support element is concerned, is fairly new. CIET came into being in 1989 as a part of the 61st Airlift Squadron stationed at Air Base No. 123 in Orléans.

Up until 1997, personnel assigned to the unit were under the CIET headquarters command at Toulouse-Francazal, but after 1 August of that year the detachment became CIET Station 01.340, attached to Airlift Squadron 02.061, more commonly called the Franche-Comté unit.

The Station has carried out missions on a regular basis for COS since that date.

Nonetheless, unlike DOS, whose operations support over 70% of COS missions, the CIET Station's primary vocation is training per-

Hercules no. 214, of the 2/61st Airlift Squadron based at Orléans. Hercules and Transalls have fundamentally different capacities. The Hercules is a high speed, long range, large capacity cargo aircraft. The Transall is better at night missions and can land almost anywhere on short runways.

Above, left:
Official insignia of the CIET Airlift Squadron (DOS/C-130).

sonnel on the C-130 Hercules. CIET devotes 75% of its time to this task, leaving 5% for CFAP (*Commandant de la Force Aérienne de Projection—Forward Air Command*) operations and the remaining 20% for COS missions.

This difference between the two units is significant; it is partly due to the lack of commonality between the airplanes used by the different units.

Currently, the Station is home to six pilots, three Navigation/Weapons Systems Officers and six mechanics.

Among the six pilots, two have Transall experience, while the other four are "only" Hercules qualified; of the mechanics only two are Transall qualified.

The pilots are without a doubt the most

experienced of all Hercules drivers, with between 2,000 and 8,000 flight hours each, and most of the unit's current members took part in the 1993 Sarajevo airlift. Part of their job is training the others.

To become unit members, applicants must be active duty, have a proven capacity for training pilots and be offered a position by the cadre of the unit.

The main talent pool for prospective CIET pilots is the Franche-Comté unit.

Like Toulouse-based DOS, CIET has no organic aircraft, but has access to two Hercules for missions and training.

The unit commander explains, "While the unit has no C-130s of its own, we can pick from the available pool of aircraft.

Besides, being a part of Franche-Comté helps because we can get additional crews if needed."

Franche-Comté has 14 Hercules C-130s, three of which have armored cockpits and anti-missile decoy flare dispensers, installed for the Sarajevo missions.

The command has elongated C-130 H-30s, with an extra two ton load capacity, and C-130 Hs.

COS-delegated missions include deep penetration air-landing of primary and secondary Special Forces personnel, HAHO/HALO jump support for 1st RPIMa elements, the Naval Commandos and CPA 10 personnel.

They also include day and night landings, low level day and night air-drops and fuel re-supply by air-dropping 1,000 liter reservoirs.

DOS/C-130's primary objective is currently to attain the same level of capability as the DOS/160-R crews.

The unit member's are counting on on-going equipment renovation programs—making cockpit and cargo areas compatible with night vision goggles by using special gray paint, like the RAF Hercules—and developing training programs tailored to COS special operations exercises that began in 1997 for DOS/C-130.

During Operation Sunrise in Jordan, C-130 crews practiced day and night combat airlanding operations and airborne drops in the desert, a mission completed in a "special operations" setting.

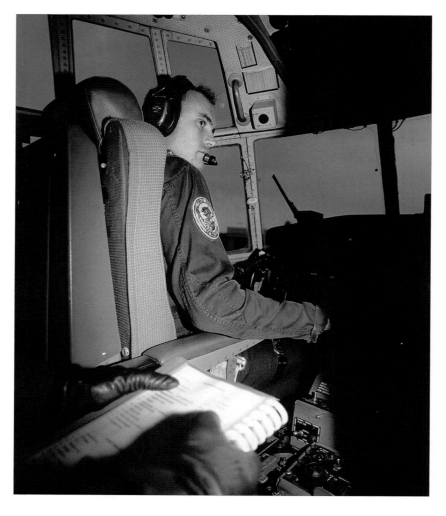

Beginning in the early fall of 1999 the Hercules crews will be working on an almost daily basis with CPA 10, who are leaving the Apt headquarters for the more practical location at Orléans, where they will share the base with their transport assets.

At left, above:
Pilots and a navigator perform pre-flight procedures at the Orléans CIET base. This recently created unit is being called on to carry out ever increasing numbers of COS missions.

At left, center:
A CIET pilot during an April 1995 flight to Sarajevo. The great majority of Hercules crews have experienced flying in a combat environment, with many hours of flying amassed over Bosnia. Here the pilot is wearing a flak jacket.

At left, below:
The CIET post commander is a C-130 pilot whose mission is to get his pilots to attain the same level of proficiency as the DOS COS-160R crews stationed at Toulouse.

Above and at right:
A series of photos taken from inside and outside of a CIET Hercules during an assault landing on a strip in the Jordanian desert in 1997. Note the tracks made by landing gear. One CIET mission is to provide long range transportation and air landings for COS elite units.

SPECIAL FORCES HELICOPTER SQUADRON

(Escadrille des Hélicoptères Spéciaux)

EHS

The French Special Forces Helicopter Squadron, known as EHS (Escadrille des Hélicoptères Spéciaux), was transferred in September 1998 to Air Base 120 in Cazaux, France. The unit, originally attached to the 5/67th Alpilles stationed in Aix-en-Provence, is now part of the 1/67th in the Pyrenees mountains. The experience of the two COS pilots who were sent over with the unit has come in handy in training other crews over a period of a few months, including three pilots and five crew mechanics.

I N THE LONG RUN, the aim of EHS is to have available five pilots and a core of five crew mechanics qualified to carry out all COS missions.

To accomplish this goal, the former Aix-en-Provence pilots will be teaching the new ones the various COS procedures, in other words RESCO (*Recherche et Sauvetage au Combat*— Combat Search and Rescue) techniques that have been adapted to COS missions. Captain Jean-Pierre Rougé observes, "The basic job is like RESCO. Although you have no signal to guide you like in a RESCO operation, you still

Preceding pages.
Two EHS (Escadrille des Hélicoptères Spéciaux— Special Forces Helicopter Squadron) helicopters, a Puma with cannon and a Super Puma, in flight over the Mediterranean. The photo was taken while they still were part of the Alpilles 5/67th Helicopter Squadron in Aix-en-Provence. Note the nonofficial EHS insignia superimposed on the photo.

Opposite.
EHS currently has no Super Pumas in the unit, shown here in flight over the Aix region in France. This helicopter is still in service in the French Air Force and could be allocated to the unit for a mission, if needed. The Puma is equipped with a "Chlio" thermal camera, while the super Puma has the "Saphir 2000". Both are suitable for night reconnaissance and intelligence gathering operations.

Above.
Air Force SA 330 Ba SAR Pumas attached to the Pyrénées squadron—and therefore EHS Pumas as well—all have hover coupler piloting systems, enabling pilots to hover safely at night.

Opposite.
Helicopter squadrons designated for COS use enjoy the advantage of access to other squadrons' bases and equipment. Thus, EHS can rely on 1/67th Pyrénées and DAOS on 5e RHC.

Below.
With the arrival of Alpilles Squadron Super-Pumas to EHS in 1995, the special operations concept is now truly being implemented.

go to a pre-determined rendezvous point and your mission is just as difficult because you're operating in hostile territory." The Captain modestly neglects to point out that COS missions come up without any warning and are totally covert, that they are solo missions conducted under radio silence, and that they take place at night with only minimal support.

All RESCO missions use two Pumas and massive support elements, fighter-bomber cover and an AWACS for surveillance and navigation. Rendezvous points are arbitrary. On the contrary, COS operations have very precise rendezvous points set at the beginning of the mission. This can be on a rooftop, a ship or in a wooded copse somewhere. The difference between RESCO and EHS crews is that the former use CFAP/RESCO

(*CFAP: Commandement de Force Aérienne de Projection*—Air Reaction Force Command) standard operating procedures, while the latter follow Special Operations procedures. This means that their area of flight coverage is much wider, even for exercises. EHS crews use the same procedure for training missions that they do during real operations. For a pilot to be considered operational at EHS, six months of closely supervised training are necessary. In addition, the pilot will have already completed a minimum of 1,500 helicopter flight hours, the equivalent of six years in rotor aircraft. Average flight time boast-ed by EHS pilots currently is between six and fifteen years of helicopter flying. Additional required qualifications include 70 to 100 hours of night vision goggles flying, 300 to 400 night flying hours, night-time and day-

HOVER COUPLERS

Hover couplers, known as CVS (*Coupleur de Vol Stationnaire*) in French, were developed in the eighties for use in helicopters. This system allows a pilot using night vision goggles to approach a point at night straight on and horizontally, and then to remain stationary at a set altitude without having to continually adjust to maintain position. The apparatus makes night operations easier and less risky. CVS was initially installed in Super-Frelons, then Pumas, Lynxs and Panthers. Pyrénées-based pilots complete two CVS missions per month. The best evidence of the utility of the CVS system is illustrated by Futura maritime insertions of naval commandos and their gear, and by night airdrops and grappe insertions and extractions.

On left.
From top, different aspects of EHS owned SA 330 SAR Puma equipment. Pilots' seats with hover coupler system at bottom of photo, the Chlio camera screen, seen from the navigator/mechanic's seat in the hold and the Chlio thermal camera positioned beneath the helicopter.

Opposite.
CPA 10 commandos rappelling from an Alpilles Super-Puma in 1998, near Sainte Victoire mountain in France. These helicopters are much more versatile than the Pumas. Circling above, a Puma provides security with 20 mm cannon.

FROM ALPILLES TO PYRÉNÉES: NEW UNIT, SAME MISSION

EHS originates with the 5/67th helicopter squadron based in Aix-en-Provence. This unit got the mission in 1989 of handling SAR (Search and Rescue) operations for the Isères, Orange and Salon-en-Provence air bases. In October 1993, one year after the establishment of COS, the French Air Force Staff decided to create a helicopter unit within the squadron specifically for special operations, calling it EHS. The new unit had five pilots, all professional military, two of which were also instructors for Puma, Super Puma and Fennec helicopters. All the pilots had impressive experience on various helicopters, boasting a minimum of 2,500 hours of flight time, with the oldest hand totaling some 4,700 hours by September 1997. There were also four flight mechanics in the original group.

In September 1995, the arrival of the Super Pumas into the squadron greatly enhanced training possibilities for the special forces context, thanks to the new equipment's superior capabilities. EHS stayed a part of the Alpilles group through 1997. At the time, the 5/67th, whose principal mission is SAR and RESCO operations—they did missions out of Brindisi, Italy—had 23 pilots, 10 flight mechanics, 12 jumper/divers and 40 ground mechanics. The unit had one Super Puma—reserved for special operations—two SAR and RESCO adaptable Pumas, one standard Puma with mounts for 20 mm cannon and 5 Fennecs used for training and ferrying missions.

The first COS operation that EHS participated in was operation Turquoise in Rwanda. The Pumas that went were used for SAR missions and mortar fire control. In another operation, an entire crew was sent to Brindisi to support air op-erations over Bosnia. This was followed by a civilian evacuation mission out of Albania in March, 1997. In addition to these deployments, EHS crews took part regularly in COS exercises and trained with elite units in France and abroad; in Senegal in 1995, and in Oman in 1996 during the Amitiés exercises.

In August 1998, EHS was transferred to Cazaux to join the 1/67th Helicopter Squadron. The unit was operation-al on 1 September 1998.

light landings on French Navy vessels (from 100 to 1,000 hours for certain pilots), hover coupler training, cannon fire exercises, night SAMAR (*Sauvetage en Mer*—Maritime Rescue Operations), Search and Rescue, CSAR and for some pilots, time spent as trainers on various helicopters.

Because EHS is officially attached to the Pyrenees group, it is not considered a separate organization. Nonetheless, it is a unit apart with two roles to play, one for COS and the other for the 1/67th, to whom the unit's Pumas are available for missions.

The unit moved its base from Aix-en-Provence for a variety of reasons. In the first place, at Cazaux there is a special IFR platform for liftoffs and landings in all weather, unlike in Aix. Furthermore, EHS is expected to operate in a maritime environment; at Aix the unit was twenty minutes flight time from the Atlantic, whereas Cazaux is two minutes away and also has a small lake within 500 meters from the units hangar where it can train. At Cazaux, there is greater equipment commonality, with 8 CVS (*Coupleur de Vol Stationnaire*—hover coupler) equipped Pumas. This is an improvement over the Aix setup, which involved prepping three different types of helicopters for missions. Target illumination and 20 mm cannon training is easier at Cazaux where ranges exist on the base itself or a step away at Capitieux. Cazaux is also closer, or at least provides easier access to, other elite unit bases including the 1st RPI-Ma at Bayonne, the Naval commando unit at Lorient and the Air commandos at Orléans.

The most important factor influencing the change, however, is that the runway at Cazaux corresponds to NATO norms, meaning that cargo planes can

Above.
In the long term, COS could decide to make EHS the lead helicopter unit for maritime operations. The unit's tradition of sea duty lends itself particularly well to the types of low altitude night missions in all types of weather that this role would entail.

146

use it. At Aix, the fact that the runway was too short meant that any airlifting of helicopters had to go through the field at Istres. Now loading of the unit's equipment into cargo holds of the big lifters is easier and quicker. As it stands, to move a Puma, COS reckons one day is needed for disassembly, one day to load it into a C-130 or C-160, the in-flight time and another day for reassembly.

EHS is capable of carrying out RESCO missions, or can be ordered to Brindisi3, in addition to its COS operations. However, unlike other RESCO crews, EHS is mission capable for COS special operations in a specified zone, in concert with another special unit, like EOS (*Escadrille des Opérations spéciales*—Special Operations Squadron for land-based missions).

EHS crews support elite unit operations using outfits like CPA 10 or the airborne RPIMa troops during fast-roping, *grappe* and rappelling, and static line paratroop drops. For combat divers like the Hubert commando unit, GCMC (Lorient) or the GIGN (GSIGN) *Gendarmerie* units, there are the CTM (*Contre-terrorisme Maritime*—Maritime Counter Terrorism) exercises. All of these operations are conducted by day or by night. COS wants EHS to become a naval operations oriented unit, mostly because its helicopters have hover couplers, a device that enables crews to safely perform flight missions at a balanced hover less than 20 meters above the ocean surface even at zero visibility. Apart from French Navy helicopters, only EHS Pumas have this capability.

With EHS set apart from EOS by virtue of its sea operations orientation, COS can pick the right tool from its pool of elite units for whatever job comes up.

"DEUXIEME CERCLE" UNITS

When COS was established in 1992, the command signed a protocol with the French Gendarmerie institutionalizing cooperation with that organization's special operations unit, the GSIGN (*Groupement de Sécurité et d'Intervention de la Gendarmerie Nationale*).

THE ROLE OF THE GENDARMERIE liaison officer that is assigned to COS—currently a former GIGN major who took part in the 1994 Marseilles airport incident—is to maintain the link between that organization, GSIGN units and the Directorate General of the Gendarmerie Nationale.

In this way he can be consulted in the event of GSIGN unit participation in military operations in France and abroad that require the expertise of this elite unit which is made up of the GIGN, EPIGN (*Escadron de Protection et d'Intervention de la Gendarmerie Nationale*—Gendarmerie Security and Assault Team) and GSPR (*Groupe de Sécurité de la Présidence*—Presidential Security Unit). This type of consultation led to elements of these units participating in operations in Bosnia, Rwanda and especially in the Comoros, where the mercenary Robert Denard was arrested. COS also consulted this officer during evacuations of French expatriates and for hostage crises occurring abroad.

Above and opposite. The 13th Airborne Dragons (Régiment de Dragons Parachutistes) perform long-range intelligence collection forays for French Military Intelligence. The unit has already worked with COS on different missions in Somalia, Rwanda, the Comoros archipelago, the Central African Republic and in the Congo. Unit personnel have worked in complete autonomy on missions or integrated into COS units. 13th RDP troops work as easily out of light reconnaissance vehicles like the P-4, the VLRA or the VAB as they do from the field in dug-in caches.

150

This liaison officer also has the mission of helping to select reservists for COS missions of one to six months, particularly for civil affairs in Bosnia.

With the creation of COS and the subsequent "allocation" of Special Forces units, many expected the 13th RDP (*Régiment de Dragons Parachutistes*—Dragons Airborne Regiment) to likewise be assigned to the new command structure. However, this role was to be filled by the BRGE (*Brigade de Renseignement et de Guerre Électronique*), an electronic warfare and intelligence brigade that was established at almost the same time. The Brigade, to which the 13th RDP was immediately attached, was assigned to French military intelligence (DRM). Nonetheless, the unit's members have often been attached to COS units during operations. They do not compete with these units, as is much repeated, but as a complement to them, because the RDP specialty is intelligence, not direct action.

There are other specialized units, apart from these leading Special Gendarmerie troops, coming from different branches of the armed forces to which COS can turn for specific missions. For example, 17th RGP (*Régiment de Génie Parachutiste*—Air-

Above and opposite.
One of the first major full-scale rehearsals for high altitude jumps requiring oxygen, at the Pau base in 1997. Gathered for the exercise were teams from the 11th Airborne Division, 1st RPIMa, the Hubert naval commandos, the 13th RDP and CPA 10.

152

154

Preceding page,
above and opposite.
Although different paratroop elements of the 11th DP have never yet carried out operations as part of a specialized COS combat group this type of situation is conceivable. An operation resulting from this type of participation would require a vast coordination effort between many different units. In these photos, HAHO/HALO airborne troops from the 1st RHP (Régiment Hussard de Parachutistes), the 17th RGP (Engineers) and the 9th RCP (Infantry).

155

borne Engineer Regiment) bomb disposal teams were attached to COS units deployed to Somalia in 1992. The command can also call on the services of GCP *(Groupement de Commandos Parachutistes)* parachutists, the 11th DP (Airborne), artillery spotters, engineers or specialized mountain units.

In fact, COS can select from the wide array of talent available in the French armed forces for missions, and the fact that it can illustrates its level of around-the-clock capability for special missions. This advantage is what sets them apart.

Above, at right and opposite:
Different views of GIGN personnel in action. The French Gendarmerie signed a protocol with COS authorizing the use of different GSIGN units—including GSIGN—for certain specific missions. COS can also use EPIGN (Escadron Parachutiste d'Intervention de la Gendarmerie Nationale) and GSPR in operations.

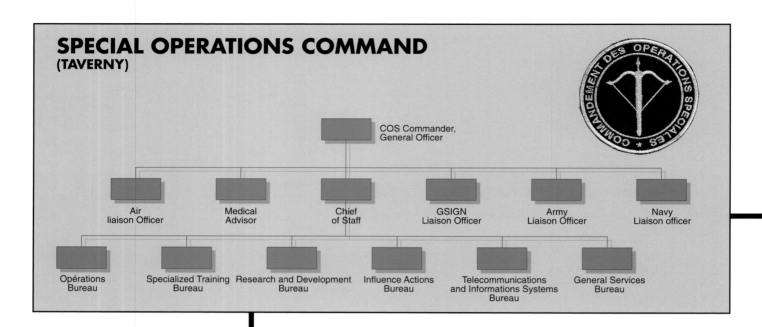

SPECIAL OPERATIONS COMMAND
(TAVERNY)

- COS Commander, General Officer
- Air liaison Officer
- Medical Advisor
- Chief of Staff
- GSIGN Liaison Officer
- Army Liaison Officer
- Navy Liaison officer
- Opérations Bureau
- Specialized Training Bureau
- Research and Development Bureau
- Influence Actions Bureau
- Telecommunications and Informations Systems Bureau
- General Services Bureau

COFUSCO

COFUSCO emerged out of the former command known as COFUSMA in 1983, which had been created in 1982 to consolidate the infantry school into the commando group. COFUSCO established itself in Lorient, and now has a headquarters of 40 people, 12 of whom are officers. It is responsible for the FUSCO—*Fusiliers et Commandos*—base, the four assault units (Jaubert, Trepel, de Pententeyo and de Monfort) and one close quarters battle group, GCMC *(Groupe de Combat en Milieu Clos)*. The command also directs the combat swimmers unit and its support vessel, the BSCN *(Batiment de support des nageurs de combat)* Poseidon and the 12 groups and companies of naval infantry. The command totals 3,000 troops spread all over the globe.

COFUSCO operates directly under the French Navy Chief of Staff and acts as its technical advisor and the general directorate authority for direct Special Forces action, Defense and the armament of light naval infantry. The five naval commando units are a part of the group of elite units available for use by COS. As such, COFUSCO is both the Navy equivalent of COS and the Navy representative to the Joint Commission on Operative Research for special operations.

EHS

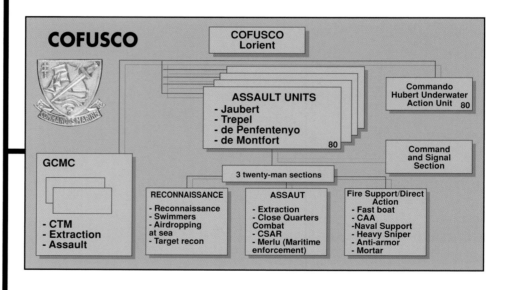

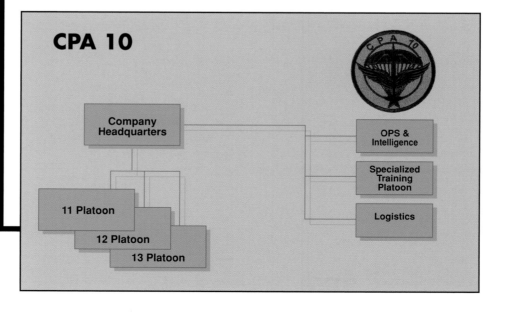

DAOS

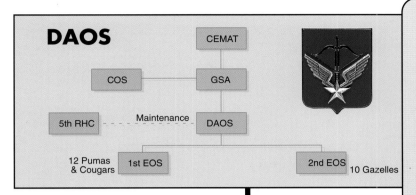

CEMAT

COS — GSA

5th RHC - - - Maintenance - - - DAOS

12 Pumas & Cougars — 1st EOS

2nd EOS — 10 Gazelles

SPECIAL AUTONOMOUS GROUP

The GSA (*Groupement Spéciale Autonome*—Special Independent Command) is an organic command that encompasses the Special Forces of the French Army, the 1st RPIMa and the ALAT(*Aviation Légère de l'Armée de Terre*—French Army Light Aviation) detachments of Special Operations DAOS (*Détachement ALAT des Opérations Spéciales*—ALAT Special Operations Detachment), the unit reserved exclusively for COS use. This detachment is responsible for maintaining the interface between Army Staff and COS. It is commanded by a General, the former 1st RPIMa commander, and is made up of three bureaus: Personnel and Training, Research and Best Use and Organization/Logistics. GSA's mission is to organize, teach, train and ensure safety for the units it commands. It defines and articulates unit logistics and equipment requirements, it manages and administers personnel and it is responsible for gearing up operational units when they are called on. GSA also assists in drawing up doctrine concerning the use of Special Forces.

1st RPIMa

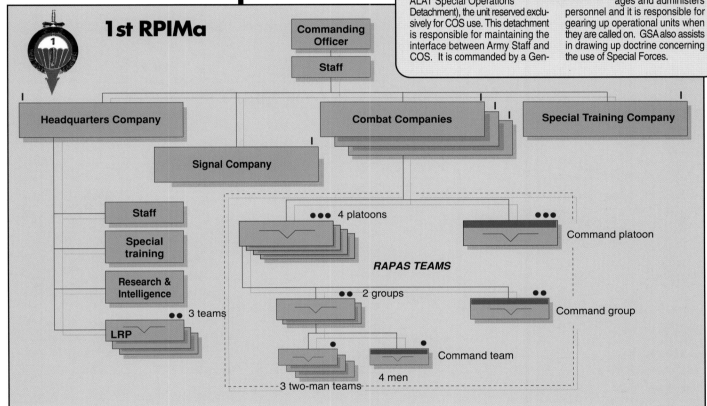

Commanding Officer

Staff

Headquarters Company

Combat Companies

Special Training Company

Signal Company

Staff

Special training

Research & Intelligence

LRP •• 3 teams

••• 4 platoons

••• Command platoon

RAPAS TEAMS

•• 2 groups

•• Command group

• • Command team

3 two-man teams — 4 men

DOS

"DEUXIEME CERCLE" UNITS

DOS-CIET

ACKNOWLEDGEMENTS

The author thanks French Air Force General Saleun, the COS Commander, Colonels de la Tousche, Doucet, Levelle and Cheroux, French Navy Captain Cavelier and Chief Commissar of the Navy Arreckx.

Also many thanks to Colonel Cheminal, Commissar Depardieu, Ensign Vachon, Lt. Cornieux, Lt. Colonel Viarouge of EICA, Major Landicheff and Captain Nöel of CPA 10.

Thanks to Navy Captain Quentin, Commander Raffin, Lt. Commander Coupry, Lieutenant Thiers and of course, Lieutenant Coupanec.

Thanks to Major Rancourt and Captain Puxeddu, of DOS in Toulouse.

Thanks to Major Bérard and Captain Dalais of the CIET station in Orléans.

Thanks also to Lt. Colonel Haefflinger and Lt. Le Gavrian of EH 05/067 Alpilles, Major Saint-Martin of the 1/61st Pyrénées, Captain Rougé and Lt. Gorce of EHS in Cazaux.

Many thanks to Gil Bourdeaux, Bruno Dallerac, Yves Debay for his excellent photographs, Laurent Ciejka, Maurice Lanlard, Jean-Michel Mailly; Wolfgang Reiss, Thierry Roger, Hervé Vernier and special thanks to Xavier Guilhou.

The author also expressly thanks the *Commando Trepel*, the Hubert combat swimmers, the commandos of CPA 10; the DOS crews, the CIET station and EHS personnel, and the personnel of the Alpilles and Pyrénées squadrons.

Translation by C.P. Mott.

BIBLIOGRAPHY

Air Actualités, n° 507, December 1997.
Armée d'aujourd'hui, n° 215.
Les Cahiers de la FED, n° 10, December 1997.
Les Cahiers de Mars, n° 140, 1994.
RAIDS n° 27, 84, 92, 97, 101, 123, 135, 137, 143, 153.

PHOTOS CREDITS

All photos by author except those cited below:
ACMAT : pages 44-45 (top and bottom)
Gérard Beaudoin : pages 16-17 (top to bottom)
Boivin/Marine nationale : page 83 (top to bottom).
Boomerang Presse : page 20 (center and bottom) page 61 (top tight, center and bottom)
CAC Systèmes : page 32 (top)
Didier Charre/ECPA : page 21 (top and bottom), page 36 (top to bottom and left to right), page 39 (top) , page 48 (bottom), page 49 (bottom), page 61 (center), page 68 (center), page 116 (bottom).
René Dalais : page 134, page 135 (top to bottom)
Yves Debay : pages 8-9 (top), page 11 (top), pages 34-35, page 37, page 41 (bottom), page 46 (top), pages 48-49 (top), page 50 (top and bottom), page 51, page 53, pages 54-55, page 61 (bottom).
DOS-CIET Toulouse : page 130 (bottom), page 131 (bottom).
Equipe ECPA : page 38 (top).
Jean-Pierre Gauthier/SIRPA-Air : page 41, page 42 (top to bottom), page 43, pages 46-47.
Giat Industries : page 33 (center and bottom)
Xavier Guilhou : pages 28- 30.
Janick Marcès/ECPA : page 22 (top to bottom), page 23 (top and bottom), page 40 (top), page 123 (bottom).
Serge-Xavier Pellizzari/ECPA : pages 10-11 (bottom).
SM Rathelot/Marine nationale : page 81 (top)
Gilles Rolle/SIRPA-Air : page 31 (bottom), pages 132-133.
Claude Savriacouty/ECPA : page 9 (bottom), page 18 (top and bottom), page 19 (bottom), page 39 (bottom), page 60 (top)
Thomson-CSF : page 33 (center and bottom)
Dominique Viola/ECPA : pages 14-15, page 20, page 30(top), page 38 (bottom), page 116 (top)
Personal collection : page 8 (bottom), page 27 (top to bottom), page 32 (bottom), pages 130-131, page 151 (top and bottom)